ST. MARY'S COLLEGE OF MARYLAND
ST. MARY'S CITY, MARYLAND 20686

Jean Heidmann

Relativistic Cosmology
An Introduction

Translated by S. and J. Mitton

With 45 Figures

Springer-Verlag Berlin Heidelberg New York 1980

Professor Jean Heidmann

Observatoire de Paris, Section d'Astrophysique
F-92190 Meudon, France

Translators:

Simon Mitton
St. Edmund's House, University of Cambridge, and

Jacqueline Mitton
New Hall, University of Cambridge,
Cambridge, CB3 8ES, England

Cover photograph from:
Compilatio Leupoldi ducatus Austriae filii de astrorum scientia decem continentis tractatus,
ed. Venetiis per Melchiorem Sessam & Petrum de Ravanis Socios 1520 (Paris Observatory Library. Photograph furnished by Hachette)

Title of the original French edition:
Introduction à la cosmologie
© by Presses Universitaires de France, Paris, 1973

ISBN 3-540-10138-1 Springer-Verlag Berlin Heidelberg New York
ISBN 0-387-10138-1 Springer-Verlag New York Heidelberg Berlin

Library of Congress Cataloging in Publication Data. Heidmann, Jean. Relativistic cosmology. Translation of Introduction à la cosmologie. Bibliography: p. Includes index. 1. Cosmology. I. Title.
QB981.H4413   523.1   80-20665

This work is subject to copyright. All rights are reserved, whether the whole or part of the material is concerned, specifically those of translation, reprinting, reuse of illustrations, broadcasting, reproduction by photocopying machine or similar means, and storage in data banks. Under § 54 of the German Copyright Law where copies are made for other than private use, a fee is payable to the publisher the amount of the fee to be determined by agreement with the publisher.

© by Springer-Verlag Berlin Heidelberg 1980
Printed in Germany

The use of registered names trademarks, etc. in this publication does not imply even in the absence of a specific statement, that such names are exempt from the relevant protective laws and regulations and therefore free for general use.

Offsetprinting and bookbinding: Beltz-Offsetdruck, 6944 Hemsbach
2153/3130-543210

For Abd-al-rahman al-Sûfi

In 964 he drew on a chart of the sky a feeble glimmer of light, which, a millennium later, was to reveal for us the realm of the galaxies.

On a moonless night, it is just possible to discern the Andromeda nebula. It is the only extragalactic object in the northern hemisphere visible to the naked eye. Abd-al-rahman al-Sūfi noticed it in the tenth century, and in 1612, Simon Marius, observing it for the first time with a telescope, described it as being like the flame of a candle seen through a thin piece of horn. In this 1667 engraving, taken from a work by Ismaël Bouillaud, it is shown as a small elliptical cloud under Andromeda's arms, in front of the fish's mouth. It was not until 1923, after the advent of large telescopes, that Edwin Hubble discovered its true nature, as a galaxy like our own, and thus opened the way for the subject of cosmology.

Its light takes two million years to reach us; it is 10^{14} times more distant than our nearest neighbour, the Moon. Nevertheless, this nebula is merely on the threshold of the extragalactic world. Galaxies a thousand times further away are already known and the distances of the quasars are perhaps larger by another factor of ten. 3 K thermal microwave radiation reaches us from even greater depths, at the edge of the cosmological horizon, which takes us to the limits of space and, by the remarkable effect of looking back in time, to the first moments of the universe, perhaps twelve thousand million years ago.

Preface to the English Edition

Since the first French edition of the book emphasized rather the solid facts of Cosmology than the detailed discussions of controversial results, relatively few revisions were necessary for the English edition. They were made early in 1979 and affected about 5% of the text. The main revisions referred to the distance scale, the distribution of galaxies, the X-ray observations of clusters, the cosmic time evolution of quasars and radiogalaxies and the 3 K radiation.

A new short bibliography presents the recent articles and the latest proceedings of Symposia; from these the reader can easily trace a more complete list of references.

I am happy to thank Professor Beiglböck for suggestions he made to improve Part II on Spaces of Constant Curvature, and Drs. S. and J. Mitton for translating the manuscript into English. I also thank with pleasure Marie-Ange Sevin for correcting the final version.

March 1980, Meudon, France *J. Heidmann*

Preface

The aim of this book is to present the fundamentals of cosmology. Its subject is the study of the universe on a grand scale:

- on a grand distance scale, since from the start, we shall be escaping the confines of our own Galaxy to explore space as far as the limits of the observable universe, some ten thousand million light years away;

- and on a grand time scale, as we shall look back into the past to the very first moments of the initial expansion, about twelve thousand million years ago.

Out to several thousand million light years, classical Euclidean geometry is good enough for tackling the various questions important to cosmology: the distance scale, the distribution of galaxies in space, the expansion of the universe, the age of the universe...

Further out in space, recession velocities reach almost the speed of light so that important relativistic effects come into play; furthermore, because light travel time is increased, our observations allow us to look back at the past history of the universe. In doing so, we enter the arena of true cosmology. We shall study its easier aspects.

To this end, we will explore spaces of constant curvature - in particular spherical space and hyperbolic space - and general relativistic models of the universe based on these spaces. Then, armed with this theoretical framework, we will present the observations of objects deep in the relativistic zone of space, radio galaxies and quasars, and the 3 K cosmic background, probably left over from the initial explosion of the universe.

The observations are presented in broad outline, without superfluous technical details, nevertheless giving the quantitative results. On theoretical matters, only the general outline of the calculations is presented, the main emphasis being placed on physical meaning. Just enough detail is given so that anyone who wishes may work through the equations as an exercise. (Where expressions such as "it may be shown that" are used, the difficulty is much greater and it is necessary to refer to one of the specialised works cited in the bibliography.)

This book is based on the course I have taught at the Faculté des Sciences in Paris and at the University of Texas. It is intended for anyone who is curious about the universe and for students who possess an appropriate background in physics and mathematics. In attempting to bring together theory and observation in the study of cosmology, one section is written for observers who wish to have sufficient theoretical knowledge to interpret their results and tackle the work of theorists. Another section is for theorists who wish to compare their calcultations with the observations and be familiar with the sources of observational results.

I am grateful to Professor V. Kourganoff for his comments which enabled me to improve the manuscript.

J. Heidmann

Contents

Part I: *The Metagalaxy out to a Distance of One Gigaparsec* 1

1. Distance Scale 3
   1.1 Cepheids 3
   1.2 H II Regions 4
   1.3 Luminosity Classes 5
   1.4 The Diameter - Luminosity Relation 6
   1.5 Groups of Galaxies 7

2. The Distribution of Galaxies in Space 9
   2.1 The Local Group 9
   2.2 The Nearby Groups 10
   2.3 The *Virgo* Cluster 11
   2.4 The *Coma* Cluster 13
   2.5 Rich Clusters 14
   2.6 Superclusters 15
   2.7 Gradients 17

3. The Expansion of the Universe 19
   3.1 The Nature of the Expansion 19
   3.2 The Value of the Hubble Constant 23

4. Nearby Intergalactic Matter 25
   4.1 Optical Observations 25
   4.2 21-cm Radio Observations 26
   4.3 X-Ray Observations 31

5. The Density of the Universe 33
   5.1 The Contribution of Galaxies 33
   5.2 The Contribution of Intergalactic Material 35
   5.3 Other Contributions 36
   5.4 The Density of the Universe 37

6. The Age of the Universe 39

Part II: *Spaces with Constant Curvature* 41

7. Locally Euclidean Spaces 43
   7.1 Natural Frame 43
   7.2 The Riemann-Christoffel Tensor 44
   7.3 Locally Euclidean Space 46
   7.4 Development 48
   7.5 Holonomy Groups 49
   7.6 Fundamental Polyhedron 50
   7.7 Representing Locally Euclidean Space in Euclidean Space 52
   7.8 The Various Types of Locally Euclidean Space 52

8. Locally Non-Euclidean Spaces 57
   8.1 First Order Representation 57
   8.2 Second Order Representation 58
   8.3 Development Along a Curve 59
   8.4 Geodesic Surfaces 62
   8.5 The Riemann-Christoffel Tensor 64
   8.6 Riemannian Curvature 65
   8.7 General Properties of Locally Non-Euclidean Spaces 68
   8.8 The Various Types of Locally Non-Euclidean Spaces 69

9. Spherical and Hyperbolic Spaces 73
   9.1 Geodesic Representation 73
   9.2 Central Representation 76
   9.3 Other Representations 78
   9.4 Appendix 83

Part III: *Model Universes* 87

10. Uniform Relativistic Model Universes 89
    10.1 The Equations of General Relativity 89
    10.2 Dingle's Equations 93
    10.3 Cosmological Solution 94
    10.4 The Robertson-Walker Metric 97
    10.5 The Friedmann Universes 99
    10.6 Radiation-Filled Universes 103

11. Theory of Observations in the Relativistic Zone 105
    11.1 Motion of Photons 105
    11.2 Spectral Ratio 106
    11.3 Travel Time of Photons 108
    11.4 Age of the Universe 110

11.5	Diameter	110
11.6	Luminosity	112
11.7	Brightness	115
11.8	Number of Observable Objects	115
11.9	Other Parameters	117

12. The Cosmological Constant 119
    12.1 The (q,λ) Diagram 120
    12.2 Evolution of the Universe 121
    12.3 Age of the Universe 124
    12.4 The Hubble Diagram 125

13. Cosmological Horizons 127
    13.1 Particle Horizon 127
    13.2 The Event Horizon 130
    13.3 The Absolute Horizon 131
    13.4 The Determination of Horizons 132

Part IV: *The Metagalaxy in the Relativistic Zone* 135

14. The Hubble Diagram for Galaxies 137

15. Distant Intergalactic Material 139
    15.1 Neutral Hydrogen 139
    15.2 Ionized Hydrogen 140

16. Radio Galaxies and Quasars 143
    16.1 Basic Data 143
    16.2 Number Counts 144
    16.3 Distribution and Luminosity Function 145
    16.4 Isotropy of Extragalactic Radio Sources 148
    16.5 Test of Closure 148

17. The Cosmic Microwave Background 151
    17.1 Description of the Cosmic Background 151
    17.2 Cosmological Interpretation 153

Numerical Constants 161
Bibliography 163
Subject Index 165

Part I
The Metagalaxy out to a Distance of One Gigaparsec

Out to a distance of one gigaparsec (Gpc)[1] the study of the metagalaxy, that is the study of the universe taken as a whole, may be done by means of classical methods. In the first place, the effects of space curvature are negligible over such "small" distances, and all the ideas and formulas available from Euclidean geometry are good enough for analysing the observations. Secondly, the relativistic effects which alter classical mechanics when the velocities of the objects being studied approach the speed of light, are themselves negligible; the greatest velocities (v) which are encountered in the metagalaxy out to one Gpc are those due to expansion, that is to say about 80,000 km s^{-1} at most, or only a quarter the speed of light. The correction term which enters most of the relativistic formulas, $\sqrt{1 - (v/c)^2}$, thus has a value of 0.96, very close to 1, and the formulas of classical mechanics are adequate.

Finally, the light from an object 1 Gpc away, has taken $3.26 \cdot 10^9$ years to reach us. The age of the universe, estimated at about $12 \cdot 10^9$ years, is relatively large compared with this light travel time. In this sense, one might say that the observations provide us with an almost instantaneous view of the metagalaxy and that it is possible to neglect the evolutionary effects: evolution of the luminosities of the galaxies, variation in the expansion velocity of the universe... If one observes a region of the metagalaxy from which the light has taken ten million years or a thousand million years to reach us, it is possible in both cases, because the general evolution of the universe is so slow, to consider that the picture obtained correspond to the present time. Evidently, there is not in fact any precise nearby limit at which the three effects of curvature, relativity and travel time, may be neglected. It all depends on the level of accuracy which is imposed; but for these purposes, one gigaparsec is a reasonable limit.

Though cosmology might be in principle the study of the universe on a grand scale, there is no need to suppose that the study of the nearby regions of the metagalaxy are of no cosmological interest. Information fundamental to cosmology is obtained in our own Galaxy, where the observation of globular clusters and the theory of stellar evolution put a lower limit on the age of the universe; the expansion is in evidence as close as a few megaparsecs. In this first part, then, we will study the properties of the nearby metagalaxy useful to cosmology.

---

[1] 1 Gpc = 10^9 pc; see *Numerical Constants*, p. 161

1. Distance Scale

The determination of the distance scale of the universe is the first fundamental task that faces an astronomer who wishes to study the universe as a whole. One needs to know the dimensions of the universe in some physical unit of distance, such as metres. These dimensions are important in themselves, but there are also other quantities which depend on the distance scale: the density, given the distribution of mass, the rate of expansion (or Hubble constant), given measured radial velocities, and so on.

Distances D in the non-relativistic zone are often expressed in terms of the *distance modulus* μ, defined in magnitudes by

$$\mu = 25 + 5 \log_{10} D \tag{1.1}$$

where D is in Mpc. μ is the difference (m - M) between the apparent magnitude m of an object at a distance D, and the magnitude M which it would have if it were situated at a distance of 10 pc. M is called the *absolute magnitude*.

1.1 Cepheids

As it is impossible to measure distances directly, we need to fall back on the use of *distance indicators*. These are objects whose intrinsic luminosities or sizes are known independently, so that their distances may be deduced by measuring their apparent luminosities and distances. The whole problem then rests on identifying and recognizing such objects. Probably the best distance indicators are the classical *Cepheid variable stars* of Population I whose periods range between 2 and 50 days. According to SANDAGE, there is a relationship:

$$f(P, M_V, B - V) = 0 \tag{1.2}$$

between period P, mean absolute visual magnitude M_V, and colour index (B - V). Measurement of the period, apparent magnitude and colour index of a Cepheid thus give its distance modulus. VAN DEN BERGH, among others, has analysed photometric measurements of Cepheids in our own Galaxy and nearby galaxies, principally the Magellanic Clouds and NGC 224. For each galaxy, the visual magnitude corrected to zero colour index $m_{V,0}$ satisfies a period-luminosity relationship with a unique slope

$$m_{v.0} = k - 2.9 \log_{10} P \qquad (1.3)$$

where k is a different constant for each galaxy. It is quite natural to suppose that this relationship also has the same *zero point* that is, the values of k would all be the same if all the galaxies were of the same distance from us.

On this hypothesis, the distance modulus μ of a galaxy is

$$\mu = k - k_G \qquad (1.4)$$

where k_G is the value of k obtained for Cepheids in our own Galaxy, using absolute magnitudes $M_{v.0}$ corrected to zero colour index in equation (1.3).

Unfortunately, the use of Cepheids is only possible for nearby galaxies, essentially the Local Group. For NGC 224, our nearest large neighbour, the visual magnitudes of the Cepheids are already down in the region of the 20th to the 22nd magnitude.

In his analysis, since revised by DE VAUCOULEURS, VAN DEN BERGH also used other distance indicators such as *novae*, *RR Lyrae* variables and *W Virginis* stars. Table 1.1 gives the adopted distance moduli, together with those obtained by SANDAGE and TAMMANN several years ago in an analysis of the same type.

Table 1.1

Galaxy	Large Cloud	Small Cloud	NGC 224	NGC 598	NGC 2403	NGC 6822	IC 1613
μ (Van den Bergh)	18.7	18.9	24.5	24.6		24.5	24.2
μ (Sandage and Tammann)	18.6	19.3		24.6	27.6	24.0	24.4

It is on these figures, as we shall see, that the distance scale of the entire universe ultimately depends.

1.2 H II Regions

Many galaxies contain numerous gaseous nebulae excited by hot stars, the H II regions, which are radiating strongly with the Hα-line from the ionized hydrogen. In general, H II regions are small, but some are huge. For NGC 598 and the Large Magellanic Cloud the diameter $\emptyset_1$, of the largest, and the average diameter of the three largest $\emptyset_3$, calculated using the distances given in Table 1.1, are listed in Table 1.2.

Table 1.2

	ϕ_1	ϕ_3
NGC 598	325	302
Large Cloud	343	224
Mean	334	263

The largest H II regions often appear as blobs of measurable size; DE VAUCOULEURS has shown, for the galaxies in Table 1.1, that their diameters vary somewhat in accordance with the luminosities or diameters of the galaxies in which they are situated.

The diameters, close to 300 pc, are not only sufficiently similar to serve as useful distance indicators, but also large enough to be measurable with large telescopes out to about 15 Mpc, much further than the distances to which Cepheids may be used. SANDAGE and TAMMANN have studied H II regions in 40 galaxies; Table 1.3 gives the distance moduli obtained by them and by DE VAUCOULEURS for four galaxies.

Table 1.3

Galaxy	NGC 247	NGC 6946	IC 10	IC 342
μ (ST)	27.7	30.1	27.4	29.5
μ (dV)	27.0	29.3	26.5	27.3

It can be seen already, at this second stage, that there are differences between the observers.

The method using H II regions, dependent on the distance scale provided by the Cepheids, extends the measurement of the universe to a much greater number of more distant galaxies, but the degree of accuracy is worse, being about 0.6 magnitude.

1.3 Luminosity Classes

VAN DEN BERGH has remarked that in spiral galaxies of type Sc, the greater the intrinsic luminosity, the greater the development of the arms. He has therefore made a systematic study of all the galaxies of types Sb, Sc and Ir which are visible on the *Palomar Sky Survey*[1] and for which the apparent magnitudes m and recession

---

[1] This atlas is composed of about a thousand plates, 6 degrees square and taken in both red and blue by the 1.2 m Schmidt telescope. It covers all the sky accessible from Mount Palomar down to magnitude 20 or 21.

velocities V are known. Their intrinsic luminosities were found by assuming a rate of expansion $H = 100$ km s^{-1} Mpc^{-1} to estimate their distances. Morphological criteria, based on the appearance of the arms for spirals, size, thickness, regularity and degree of resolution, and for irregulars based on the degree of density and fragmentation, showed that it is possible to divide the galaxies into 5 luminosity classes numbered I to V. The mean absolute photographic magnitude M_{pg} of each according to a revision made by BOTTINELLI and GOUGUENHEIM is given in Table 1.4.

Table 1.4

Class	I	II	III	IV	V
Absolute Magnitude	-20.6	-19.9	-19.0	-17.8	-14.5

The method of determining distances by luminosity class consists, therefore, of finding the class by VAN DEN BERGH's morphological criteria, which gives the absolute magnitude, then deducing the distance modulus by comparison with the apparent magnitude. Thus, VAN DEN BERGH has given a list of 935 moduli based on an expansion rate $H = 100$ km s^{-1} Mpc^{-1}. For any other value of H, from their definition, the moduli will be changed by an amount $-5 \log_{10}(H/100)$. In Table 1.1, there are two galaxies with well established classes: NGC 224 intermediate between I and II and NGC 598 of class II-III, which lead to an evaluation of H. Overall, the standard error on moduli obtained by means of luminosity classes is 0.65 mag., that is 30% of the distances. The error is greater than in the previous methods but the luminosity class method does have the advantage of enabling us to probe further in the universe, since the morphology of the arms of a giant spiral can be recognised, with the aid of a large telescope, out to a distance of some twenty megaparsecs.

1.4 The Diameter - Luminosity Relation

For a long time, astronomers have noted a reasonable correlation between the diameter and intrinsic luminosity of galaxies. HEIDMANN has shown that, if the galaxies are grouped by their different morphological types instead of being mixed together, the relationships for each type become much more precise.

If L is the intrinsic luminosity and A the linear diameter, then

$$L \propto A^q \qquad (1.5)$$

with a value of q close to 2 for ellipticals, and much larger for spirals.

When the value of the exponent q differs significantly from 2, the distance modulus μ of a galaxy may be deduced from its magnitude m and its apparent diameter a by

$$\mu = (m + 2.5 \, q \, \log_{10} a)/(1 - q/2) + K \qquad (1.6)$$

where K is a calibration constant.

The relationship for the spirals in the *Virgo* cluster, all of which are at essentially the same distance from us, gives the gradient in equation (1.5): $q = 2.6 \pm 0.1$, while the calibration constant K is obtained from the standard distance moduli in Table 1.1. This method is less precise than the preceding ones, but has enormous scope as it only requires the photometric measurement of magnitude and diameter: thus a 15th magnitude giant spiral with an apparent diameter of one minute of arc is at a distance of 150 Mpc.

Another method is based on a direct relationship found by TULLY and FISHER between the luminosity L and the maximum rotation velocity V_m of galaxies. Roughly speaking, the mass of a galaxy found from dynamics is given by $\mathcal{M} \propto A \, V_m^2$; furthermore, for spirals $\mathcal{M}/L$ is nearly constant; thus, the relation (1.5) leads to

$$L \propto V_m^{2/(1 - 1/q)} . \qquad (1.7)$$

V_m is measured from the width of the profile of the 21-cm line emitted by neutral hydrogen in galaxies. This width is equal to $2V_m \sin i$, where i is the inclination of the galaxy.

Finally, VISVANATHAN and SANDAGE have found a relation between the colour and luminosity for ellipticals which also enables distance estimates to be made.

1.5 Groups of Galaxies

With the help of the large *Catalogue of Bright Galaxies*, which lists the 4400 brightest galaxies in the sky, DE VAUCOULEURS, whose wife has helped to establish the Catalogue, has systematically researched groups of galaxies. He has discovered about fifty of them, to which subject we will return in Chapter 2.

To determine their distances, he proceeds in stages. First he determines the magnitude and the mean diameter of the first five galaxies of each group. Thanks to these averages, the relative distances of the nearest dozen groups are established through the use of secondary distance indicators, such as H II regions and luminosity classes; then these relative distances are calibrated with the aid of primary distance indicators - Cepheids, novae, RR Lyrae stars - in terms of the distances of the galaxies in the Local Group given in Table 1.1. DE VAUCOULEURS then obtains, with the aid of these calibrated averages, the distances of 40 more distant groups. Finally, he gives the distances of between 250 and 300 galaxies up to about fifteen megaparsecs away.

In conclusion, the distances of several hundred galaxies are known, principally by five different methods: the diameters of H II regions, luminosity classes, the diameter-luminosity relation, the rotation speed-luminosity relation and groups.

The distances thus reached are some ten to twenty megaparsecs. These distances rest finally on the distances of some half-dozen nearby galaxies, no more than several megaparsecs away. Above all else, the distances depend on the comparison of Cepheids with those in our own Galaxy, supposing that these stars have the same properties in all galaxies.

The methods described in this chapter are all independent of the expansion rate of the universe, which none of them uses. Only that based on luminosity classes rests on the hypothesis that the distance of a galaxy is proportional to its recession speed, but it does not in fact use a definite value for the constant of proportionality.

Work is in progress to improve the methods by which the distances of galaxies are measured, for developing them and finding new ones. Observations of the 21-cm line of atomic hydrogen seem promising in this area.

Thanks to these distances, the expansion of the universe may be established and calibrated (Chapter 3). It will suffice, then, for estimating the distances to the more distant objects, to measure their speeds of recession.

2. The Distribution of Galaxies in Space

The galaxies are the most obvious constituents of the universe. In order to determine the rate of expansion and the density of the universe, fundamental parameters affecting its evolution, it is therefore important to study the distribution of galaxies in space. We will make our exploration starting with the nearest galaxies, moving progressively deeper into space from the Local Group out to about a gigaparsec.

2.1 The Local Group

Among the several hundred or so galaxies for which we have obtained distances by the means outlined in the previous chapter, there are 27 whose distances are less than 1.3 Mpc and not any lying between 1.3 and 2.4 Mpc. There exists, then, a *Local Group* of galaxies, of which we are part, clearly separate from the other galaxies in the universe.

The Local Group includes two giant spiral galaxies: our own Galaxy and the Andromeda nebula, also known as NGC 224 or Messier 31, two average spirals: the Triangulum nebula (NGC 598 or M 33) and the Large Magellanic Cloud, eight dwarf irregular galaxies, eleven dwarf elliptical galaxies that are sparsely populated, four objects intermediate between galaxies and globular clusters, and perhaps, finally, two globular clusters stranded in deep space. The group seems quite dispersed, without any central condensation. It is perhaps composed of two sub-groups centred on NGC 224 and on our own Galaxy. It occupies a volume of $2 \cdot 1.5 \cdot 1$ Mpc^3.

Its *luminosity function*, that is the number of galaxies $N(M_{pg})$ more luminous than the absolute photographic magnitude M_{pg} - corrected for all the absorption in our Galaxy - is given at the bottom of Fig. 2.1. The Local Group contains 11 galaxies brighter than $M_{pg} = -14.5$. Adding the luminosities of the members of the Local Group, its total absolute photographic magnitude, without correction for the absorption in the interiors of the galaxies, comes to approximately -21.5, corresponding to a luminosity $L_{pg} \simeq 60 \cdot 10^9 L_\odot$. Its total mass is $650 \cdot 10^9 M_\odot$ [1], of which 70% lies in the two giants: $180 \cdot 10^9 M_\odot$ in our Galaxy and 240 to $310 \cdot 10^9 M_\odot$ in NGC 224. The ratio of mass to luminosity for the Group is then $\mathcal{M}/L_{pg} \simeq 10$. The masses given here are the "classical" masses; see the discussion at the end of 5.3.

---

[1] $L_\odot$ and $M_\odot$ are the intrinsic luminosity and the mass of the Sun respectively.

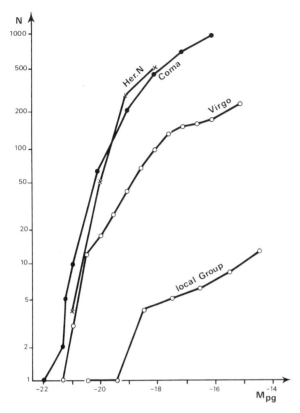

Fig. 2.1. The integral luminosity functions of the Local Group and various other clusters

2.2 The Nearby Groups

DE VAUCOULEURS has identified 14 groups of galaxies within 10 Mpc, a distance where his count is complete. The nearest to us is the group in *Sculptor*, at 2.4 Mpc; it contains six beautiful galaxies, all of type Sc, which indicates a kind of specialisation in the population. Nearly circular, its diameter in the sky is 20°, or actually 1 Mpc. It too is truly isolated.

Next comes the second group, that in *Ursa Major-Camelopardalis*, at 2.5 Mpc; quite dispersed, it occupies 2·1 Mpc, with a strong concentration around NGC 3031 to the north, and one to the south, behind the Milky Way, including the obscured galaxies *Maffei* 1 and 2.

Within the 16 Mpc limit of his investigation, DE VAUCOULEURS has counted 54 groups of galaxies. In Fig. 2.2 their positions are shown projected onto a meridian "supergalactic" plane passing through the *Virgo* cluster (see below). The Local Group is at the centre of the diagram. Apart from the shaded zone, which is obscured by the Milky Way, the count is practically complete. The figure shows that the general distribution of the groups in space is not regular: there are more groups to the

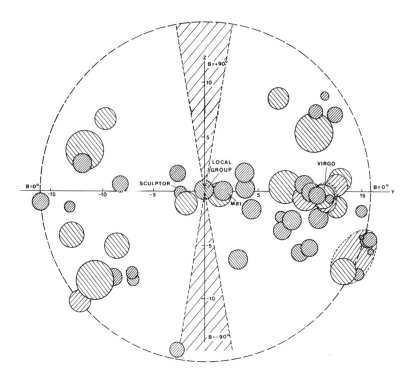

Fig. 2.2. The groups of galaxies within a distance of 16 Mpc

right and a concentration on this side towards the plane whose projection is along the horizontal diameter - it is the supergalactic plane which we shall meet again later.

On average, there is a group every 7 Mpc; their diameters range between 0.5 and 4 Mpc, with a mean of 2 Mpc. They are well separated from each other.

Their luminosity functions are only established for the brightest galaxies, down to M_{pg} = -18.7. Thus far, the nearest ones contain an average of 3 or 4 galaxies, and resemble closely the Local Group. Those more distant - the *Virgo* cluster excluded - have about 8 galaxies, perhaps on account of the tendency to separate the remote groups less well.

Not all galaxies belong to a group. According to DE VAUCOULEURS, 10 to 20% of galaxies are isolated in space.

2.3 The *Virgo* Cluster

The distribution of the 4400 galaxies in the *Catalogue of Bright Galaxies* shows up a large cloud of them, revealing the presence of a new kind of object. The first large survey in this region was carried out by AMES in 1930; if space were uniformly populated with galaxies, $\log_{10} n$, where n is the number of galaxies brighter than magnitude m_{pg}, should increase as 0.6 m_{pg}, whereas the number count in the zone in question (12 to 13 hours R.A. and 5° to 20° declination) shows an excess of several hundred galaxies, cutting off around m_{pg} = 14.5. We thus have an important cluster which is well-defined, not only in angular size in the sky, but in its range of distance in space. This is the *Virgo* cluster.

Luminosity function - In order to obtain the luminosity function of the cluster, it is first necessary to estimate the background due to the general population of the universe, in order to subtract it. For this purpose, HOLMBERG used the *Catalogue* by SHAPLEY and AMES - the first large catalogue of galaxies, produced in 1932 and containing the 1300 brightest objects. He obtained for the number of galaxies per square degree n brighter than magnitude m_{pg}, excluding the *Virgo* zone

$$\log_{10} n = 0.6 \, m_{pg} - 8.9 \tag{2.1}$$

valid down to m_{pg} = 12. Taking away this background from his observations in the *Virgo* zone, HOLMBERG obtained the luminosity function shown in Fig. 2.1, plotted for a distance modulus μ = 30.5 for the cluster (cf. below), all the absorption in our own Galaxy being taken into account. The points around $M_{pg} \simeq -15$ come from a study of 50 dwarf galaxies in the cluster by REAVES.

Figure 2.1 shows that the curve for the *Virgo* cluster is similar to that of the Local Group, but is displaced upwards by an amount corresponding to a factor of 20. Thus the *Virgo* cluster is twenty times more important than a group of galaxies; we are encountering a new entity in the population of the universe, containing several hundred galaxies. Its total photographic magnitude (the absorption of our Galaxy being taken into account) is 5.7 and its absolute magnitude -24.8, corresponding to

$$L_{pg} = 1.2 \cdot 10^{12} \, L_{\odot}.$$

Structure - Taking account of all the available criteria: morphological type, position, radial velocity, magnitude, diameter, DE VAUCOULEURS has drawn up a list of 212 galaxies which he considers to be definite or possible members of the *Virgo* cluster, down to m_{pg} = 13.5. The spirals and irregulars are distributed uniformly in an elliptical cloud 9° · 5°, while the ellipticals and S0 types, which account for about 30% of the population, are in a circular cloud, 12° in diameter with a strong concentration towards the centre.

Radial Velocities - The distribution of radial velocities in the two clouds are very different, one being quite flat, and the other much more spiky. Both are very large, covering a range of 2000 km s^{-1} centred on about 1000 km s^{-1}. It might be supposed that the spirals and irregulars, less massive than the ellipticals, would have a greater spread in velocities, being also more dispersed in space, as we have seen.

The diameter of the *Virgo* cluster is 12°, or 2.7 Mpc for μ = 30.5. It is comparable with the sizes of groups of galaxies. The cluster is, therefore, very much more dense.

2.4 The *Coma* Cluster

We are going to jump one step further in our exploration of the Metagalaxy, both in distance and in the importance of the objects encountered. The galaxy counts carried out by SHANE and his collaborators at Lick Observatory, to a limit of $m_{pg} \simeq 18.7$, reached galaxies 5 magnitudes fainter than DE VAUCOULEUR's, and thus ten times further out and encompassing a volume of space a thousand times greater. A million galaxies have thus been located and it is possible to construct contours of equal galaxy density, their number per square degree.

There is a more or less uniform background of about 50 galaxies per square degree against which stand out dozens of clusters with densities exceeding 100 galaxies per square degree. On such a map, the *Virgo* cluster is not visible, being too spread out. Near the north galactic pole there is a particularly rich cluster: the *Coma* cluster. This is one of the rich, compact types. It is one of the nearest and one of the most studied.

Galaxy counts in this cluster, by circular zones at increasing radii, have been made by ZWICKY, down to the magnitude limit of the Palomar 1.2 m-Schmidt telescope. The numbers decrease very quickly and finish up lost in the general background, without a precise limit. OMER and PAGE have counted 700 galaxies in a radius of 100 arc min down to m_{pg} = 18.8. This number increases very rapidly with magnitude limit: 2500 to m_{pg} = 19.3 in the same radius, background excluded. The counts taken to the faintest limits have been made by NOONAN, with the Hale 5-m telescope, down to m_{pg} = 22.4; just within a 5-arc min radius the cluster contains 100 galaxies, of which the faintest are dwarfs with absolute magnitudes around -12.

The distribution in magnitude has been determined within zones of fixed radius; that done by ABELL is reproduced in Fig. 2.1; it is similar in shape to the *Virgo* curve, but is shifted vertically by a factor of 3.

Thus, in a diameter of 100 arc min, or 2.7 Mpc (using the distance calculated below), which is the same as the diameter of the *Virgo* cluster, the *Coma* cluster is still three times richer, equivalent to sixty times the Local Group.

Combining the radial distribution to a given magnitude with the magnitude distribution to a given radius, HEIDMANN has deduced the cluster's luminosity as a function of the distance from the centre. The apparent photographic luminosity l, contained in a circle of radius r is proportional to r, $l = c_0 r$, out to $r \simeq 70$ arc min. From this, the total magnitude of the cluster within a radius $r = 50$ arc min is deduced as m_{pg} = 9. Using a distance modulus μ = 35, M_{pg} = -26. That is, $L_{pg} = 3.5 \cdot 10^{12} L_\odot$. For r = 100 arc min, the "official" limit of the cluster, L_{pg} reaches $6 \cdot 10^{12} L_\odot$, five times the value for the *Virgo* cluster.

The surface brightness of the cluster varies as $1/r$, which allows the centre to be located to within an arc min. The brightness is strongly peaked towards the central region, but according to NOONAN, is flattening out for $r < 3$ or 5 arc min.

In space, if one assumes spherical symmetry, this law of brightness leads to an emissivity E(r), that is, luminosity per unit volume, varying as $1/r^2$. This variation is the same as the variation of density in EMDEN's self-gravitating, isothermal sphere of gas, which is discussed in Chapter 5. It indicates that a certain state of relaxation has been attained by the cluster, probably when the galaxies were no more than clouds of protogalactic gas.

Two hundred radial velocities for member galaxies have been published. The average recession speed is 6960 km s^{-1} with respect to our Galaxy, with a dispersion of 896 km s^{-1}. With the expansion rate estimated in the next chapter, the distance to the *Coma* cluster is of the order of 100 Mpc, or $\mu = 35$.

The cluster is predominantly populated by elliptical galaxies, indicating here also, a specialised population.

2.5 Rich Clusters

In the *Coma* cluster, we have met an example of a rich cluster of galaxies and we will now go on to tackle the problems concerned with the existence and distribution of these gigantic objects. It will take us even further in the exploration of space.

ABELL has systematically researched rich clusters on the *Palomar Sky Survey* plates taken in red light, down to magnitude $m_{pg} = 21$, that is three times more distant than SHANE's surveys. For this research, he chose distance independent criteria, which is very important for obtaining a systematic survey. He has thus drawn up a catalogue of 2700 clusters.

The distribution in richness is given in Table 2.1 according to the number of galaxies R in each cluster brighter than $m_3 + 2$, for the clusters well clear of galactic absorption, where m_3 is the magnitude of the third galaxy in each cluster in order of decreasing brightness.

Table 2.1

R (galaxies)	50	80	130	200	300
Number of clusters	1224	383	68	6	1

The *Coma* cluster falls in the second category; clusters three times richer exist, but the richer the clusters, the smaller the number known.

Not all clusters have the relatively regular shape of the *Coma* cluster. ROOD and his collaborators have come up with the basis of a morphological classification for the nearest 111 of ABELL's clusters, based on the distribution of the ten brightest galaxies; this distribution may be spherical, linear, planar or irregular. MORGAN and WHITE have introduced other parameters: contrast, compactness and concentration.

ABELL found that the distribution with distance is given by the expression:

$$\log_{10} N = 0.6\, m + \text{constant}$$

where N is the number of clusters having m_{10} less than or equal to m, with an accuracy of 50%. Thus, space is, roughly speaking, populated uniformly with rich clusters to within a factor of 2, down to the limit of ABELL's survey, which is estimated to correspond to a recession velocity of 60,000 km s^{-1}, a distance of around 800 Mpc.

The average separation of rich clusters can then be calculated: there is one cluster every 55 Mpc, that is one per 175,000 Mpc^3.

The large scale distribution of ABELL's clusters on the celestial sphere seems fairly isotropic. OORT has studied the distribution for the more distant clusters - consequently the most numerous, and those which are most interesting in large expanses of space - in the distance range 600 to 800 Mpc. The numbers in the quadrants of the 30° galactic polar caps are on average 54, and there does not seem to be any great difference between them except for one which contains 37. If this difference is not due to stronger galactic absorption, it shows that on a distance scale of the order of a gigaparsec, there are variations in the population density of the universe not larger than a factor of 2. Without inferring that the universe is homogeneous and isotropic, one could say that on a scale of a gigaparsec the universe is populated homogeneously and isotropically, to within a factor of 2.

Another important catalogue of clusters has been constructed by ZWICKY and his collaborators from the *Palomar Sky Survey*, north of declination -3°. As well as listing individually 30,000 galaxies brighter than m_{pg} = 15.7, he gives the characteristics of 9700 clusters of galaxies.

The apparent diameters of ZWICKY's clusters, on average, vary as the inverse of distance, corresponding to a linear diameter of 10 Mpc. This value is larger than those of ABELL's clusters, since here, some clusters separated by ABELL, are grouped by ZWICKY into *open* clusters.

2.6 Superclusters

It appears uncertain, whether the rich clusters we have discussed are the largest entities in the population of the universe. The map shown in Fig. 2.3 shows SHANE's number contours in the region of *Hercules*. Obviously, we are seeing here a cluster of clusters, the members of which are genuine rich clusters. The luminosity function of the North component of *Hercules* has been constructed by HEIDMANN, and is shown in Fig. 2.1, for a radius of 1.6 Mpc, comparable with the dimensions of the *Virgo* cluster. *Hercules North* is as rich as the *Coma* cluster. Its total magnitude is M_{pg} = -26.0 and its total luminosity, $L_{pg} = 3.5 \cdot 10^{12}\, L_\odot$. This rich cluster, unlike that in *Coma* which is mainly composed of ellipticals, principally contains spirals. The total size of the three clusters is 20 Mpc.

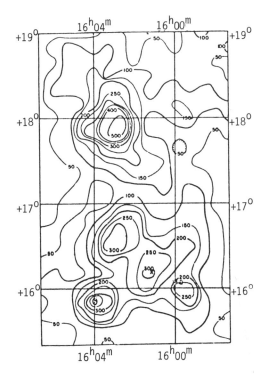

Fig. 2.3. The supercluster in *Hercules*

Another indication of the existence of groupings with such vast dimensions is provided by the Supergalaxy, or Local Supercluster; according to DE VAUCOULEURS, our Local Group forms part of a vast system, centred on the *Virgo* cluster, and having a radius of 15 Mpc. It is a more or less flattened system, whose equatorial plane defines the supergalactic plane which was discussed earlier. Returning to Fig. 2.2, we see that the Local Group distinctly appears to be situated on the borders of the system, seen here edge on. A little further from the *Virgo* cluster than our Group, the *Sculptor* group seems to mark the limits of the Local Supercluster, while at even greater distances, a universal background of groups appears, clearly separated from the Supercluster.

A general study of clusters of clusters has been made from his catalogue by ABELL, showing the existence of "aggregations" of clusters around 60 Mpc in diameter. ABELL found about fifty superclusters containing an average of ten clusters each. The biggest contains 29 clusters; this is just the count of rich clusters, remember, for in addition it contains dozens of smaller clusters.

Superclusters are, then, tremendously rich aggregations, each equivalent to around five times the *Coma* cluster, leading to a total luminosity of $L_{pg} \simeq 30 \cdot 10^{12} \, L_\odot$.

DE VAUCOULEURS and PEEBLES have analysed the angular correlations in SHANE's counts and found aggregations of the order of 50 Mpc in size. In analysing the distant clusters in ZWICKY's catalogue, KARATCHENTSEV obtained the same dimensions with 5 to 10 clusters per supercluster, and one supercluster every 10^7 Mpc^3.

Not all clusters are members of superclusters. According to PEEBLES, less than 10% of them would be in a cluster of clusters, whereas ROOD found a figure of 40%, and that most clusters are in pairs or triplets; according to him, clusters of clusters of clusters probably do not exist. In fact, the situation is not always clear cut; the edges of superclusters are not well-defined and a statistical analysis by KIANG and SASLAW indicates that, with a mean diameter of 75 Mpc, they may touch or even encroach each other.

2.7 Gradients

We have surveyed space almost as far out as a gigaparsec, and we have identified the largest known objects, the superclusters. Are there any objects even larger still, clusters of the third order? If they do exist, with the distance limits on the observation of galaxies, it is not possible to observe more than one or two, or parts of several of them. To disclose their presence, it is necessary to study the distribution of galaxies on a very grand scale, and to try to reveal, eventually, concentration gradients: either angular gradients on the celestial sphere or gradients in distance in space.

We have seen that the distribution of ABELL's clusters shows the possible existence of gradients of a factor of 2 over angles of the order of a radian. Further information is supplied by the study of the general distribution of galaxies, located one by one by position and magnitude over vast areas of sky, and to great distances. This work requires precise, individual, nebular photometry in order to obtain the contours of number counts N(m) giving the total number of galaxies per square degree of the sky, brighter than magnitude m, for various values of m.

The deepest surveys of this type have been made by SHAPLEY. Around the south galactic pole, 80 000 galaxies have been measured; the counts down to m_{pg} = 18.5 where the measurements become incomplete, lead to

$$\log_{10} N(m_{pg}) = (0.59 \pm 0.01) m_{pg} - 9.1 \pm 0.02 \tag{2.2}$$

which is very close to the relation given in (2.1). Thus, in this area of sky, there is not any indication of the approach of the edge or centre of a vast system.

On the other hand, across this section, N(18) varies by a factor of 2 over an arc of 40°, indicating a slight gradient. Around the north galactic pole, with a survey of 95,000 galaxies, SHAPLEY obtained comparable results with, perhaps, a similar angular gradient.

DE VAUCOULEURS did not detect any difference between the north and the south, a conclusion confirmed by his examination of SHANE's counts, and the older but deeper ones done by HUBBLE. There is not, then, evidence for any difference exceeding several percent between the mean densities in two cone-shaped volumes of space extending in opposing directions, towards the north and the south, each encompassing between two and three steradians, and extending for 250 Mpc, the centroids of the two being separated by 350 Mpc.

The general conservative conclusion is that on the scale of a gigaparsec, the universe is uniformly populated with galaxies, within a factor of about 2, implying that above the scale of superclusters, which reach perhaps 100 Mpc, gradients probably exist corresponding to a factor of 2 at most in the counts of galaxies.

To go further, it will be necessary to make detailed observations with 1.2 m Schmidt telescopes, or large, wide-field *Ritchey-Chretien* telescopes, such as the 3.6 m at the European Southern Observatory. (The 5 m-telescope at Mount Palomar has too narrow a field).

One of the most distant clusters that has been observed is that surrounding the radio source *Bootes* A; it contains 60 galaxies brighter than visual magnitude 21 within a radius of 1.5 minutes of arc, and its recession speed is 140,000 km s^{-1}, but its fortunate discovery was due to the presence of the radio source. With new Kodak 087 plates, it is possible to reach down almost to 25th magnitude, which will enable astronomers to find even more distant clusters. Extrapolating to m_{pg} = 21, the relation (2.2) leads to one galaxy every square arc min, 100 million over the whole sky. But we are beginning to enter the relativistic zone, the next stage reached by means of the radio sources.

We will end with what might be called the cellular structure of the universe. EINASTO has remarked that clusters have a tendency to be organised in lines, while TARENGHI discovered galaxies arranged in sheets and TULLY showed the existence of regions of space completely devoid of galaxies. Schematically, the way the universe is filled with galaxies could be thought of in terms of blocks. Inside the cubes there is nothing. There are galaxies on the faces. Along the lines where the faces intersect there are more galaxies, and at the corners, even more. In fact, these cells are not properly formed, but have dimensions of the order of 100 Mpc.

3. The Expansion of the Universe

The first great cosmological discovery was surely the expansion of the universe, at least from the standpoint of astronomers working in the late twentieth century. Around 1930, HUBBLE discovered that the whole world of "island universes" beyond our own Galaxy is expanding. All the galaxies rush away from us with speeds up to 1000 km s^{-1} for the nearby ones, 10,000 km s^{-1} for the more distant and 100,000 km s^{-1} or more for the furthest.

Objects of considerable mass, 10^{44} g, are travelling at speeds comparable with that of light. And not just some, but all of them. Nor are they moving at random, but according to an elegant scheme, simply that the further away they are, the faster they go.

This universal phenomenon, of considerable intrinsic interest, also poses questions of prime importance: extrapolating back into the past, one arrives at some origin for the universe, at its "creation"; extrapolating into the future, do we approach little by little empty nothingness?

This discovery, these questions and their answers were, in fact, contained already in the memoirs of FRIEDMANN, based on EINSTEIN's General Relativity. Disregarded for almost ten years, this theoretical work is now, half a century later, still in favour, a remarkable fact, indeed.

The phenomenon of the expansion of the universe breaks down into two aspects:
- the nature of the expansion: are the recession speeds of the galaxies really proportional to their distances?
- the rate of expansion: what is the value of the constant of proportionality?

These two problems, though having fundamental cosmological importance, are solved by observations in the "classical" region, in the nearby universe, less than a gigaparsec away.

3.1 The Nature of the Expansion

The law of the expansion of the universe relates radial velocity with distance, and since we are only interested in the form of the law, it is sufficient to use relative distances, rather than distances expressed in units such as centimetres.

Measurement of velocities - The radial velocities v are measured by the shift $\Delta\lambda$ of a spectral line and the classical Doppler-Fizeau law

$$v = c\, \Delta\lambda/\lambda \tag{3.1}$$

where λ is the laboratory wavelength of the line and c the velocity of light. This measurement does not pose any major problems, particularly for values of v which are relatively large. On the other hand, the reference frame with respect to which these velocities are measured has a particular importance because it may introduce into the study of the expansion various unwanted motions:

1. The movement of the Earth with respect to the Sun;

2. The movement of the Sun towards the apex with respect to the *local reference frame* used in galactic astronomy;

3. The movement of the local reference frame with respect to our Galaxy; that is, what is commonly called the orbital motion of the Sun about the galactic centre, or the galactic rotation of the Sun. Still relatively ill-determined, its velocity is of the order of 250 km s^{-1} in the direction of galactic coordinates $l = 90°$, $b = 0°$;

4. The motion of the Galaxy, relative to the Local Group of galaxies. Using the radial velocities of thirteen galaxies in the Group, DE VAUCOULEURS and PETERS have found that the Sun moves at 315 km s^{-1} in the direction $l = 95°$, $b = -8°$ with respect to the Local Group. Our Galaxy moves, then, at about 80 km s^{-1} with respect to the Group, a normal and relatively small velocity;

5. The motion of the Local Group with respect to the nearby Metagalaxy, to within several megaparsecs.

In order to improve the study of the expansion of the universe it is interesting to adopt a reference frame linked to a portion of space, sufficiently large to be free of the individual movements of the galaxies as much as possible, and not so large that the general expansion is not negligible. With respect to the galaxies not belonging to the Local Group, but situated less than 6 Mpc away, GOUGUENHEIM obtained (250 ± 40) km s^{-1} towards $l = 130°$ and $b = 9°$ for the motion of the Sun.

The Local Group is then a fairly slow mover relative to the nearby Metagalaxy and the Galaxy, too. All in all the most important unwanted velocity results from the galactic rotation and it is sufficient to adopt the Local Group as a reference frame, a system in which our Sun has a velocity of very nearly 300 km s^{-1} towards $l = 90°$, $b = 0°$. On the other hand, with respect to the more distant Metagalaxy, between 50 and 100 Mpc away, RUBIN and FORD have shown that the Local Group is moving at 450 km s^{-1} towards $l = 163°$, $b = -11°$; in this motion, our Galaxy moves almost edge-on in the direction of the anticentre.

Measurement of relative distances - In order to obtain the form of the law of expansion with the greatest possible accuracy, it is necessary to observe galaxies out to great distances, both to extend the law over the largest distance range and also to ensure that the expansion velocities are clearly greater than the random individual motions of the galaxies. At such distances, much greater than that of the *Virgo* cluster, the distance indicators which were discussed in Chapter 1 are not usable, except that based on the diameter-luminosity relation which is too recent a discovery to have been applied in this area.

The major statistical study of HUMASON, MAYALL and SANDAGE rests on the use of apparent magnitudes. The principle is the following: if the expansion velocity v is related to the distance D by

$$v = HD \qquad (3.2)$$

where H is a constant of proportionality, *Hubble's constant*, then, since there is a relation between the apparent bolometric magnitude m_{bol} and the absolute bolometric magnitude M_{bol}

$$m_{bol} - M_{bol} = 25 + 5 \log_{10} D, \qquad (3.3)$$

the distance D being in Mpc, one obtains

$$\log_{10} v = 0.2 \, m_{bol} - (5 - \log_{10} H + 0.2 \, M_{bol}), \qquad (3.4)$$

valid, remember, for Euclidian space and velocities v, small relative to c.

If we suppose that the galaxies all have the same absolute magnitude M_{bol}, a diagram giving $\log_{10} v$ as a function of m_{bol}, called the *Hubble diagram*, should give a straight line of slope 0.2 for a linear expansion (3.2).

The difficulty with the method lies in the corrections which have to be made to the magnitudes, most of all the *K-correction*, which converts the measured magnitude, m_{pg} for example, into m_{bol}; it is only since the recent work of OKE and SANDAGE, and PENCE, that this correction has been well established.

In their major work on the expansion, HUMASON, MAYALL and SANDAGE gave the radial velocities v of 600 galaxies, relative to the Local Group, and their corrected magnitudes. Fairly distant measurements are involved as magnitudes of 20 and recession speeds of 60,000 km s^{-1} are reached. For 474 galaxies not belonging to clusters, a fairly significant dispersion is observed. In magnitude it is mainly for the nearby galaxies, being due to the fact that the ones having low intrinsic luminosities are more easily observed locally than they are at greater distances. Dispersion also occurs in velocity, arising from random motions of galaxies, perhaps reaching 200 km s^{-1}, and also from errors of measurement, perhaps as great as 100 km s^{-1}.

Statistical analysis leads to a straight line with a slope 0.199 ± 0.005. Thus, by virtue of equations (3.2), (3.3) and (3.4), the recession velocity is proportional to distance, with a restriction, however, that there is no uniform intergalactic absorption which would make a nonlinear expansion appear linear; this possibility would be such a coincidence and have such a geocentric character, that it is improbable.

The statistical analysis also shows a difference between the galaxies of the north and south galactic hemispheres. This is not a matter of anisotropy in the expansion of the universe, but simply a spurious effect introduced by the lights of Los Angeles; constant vigilance iş needed or an effect due to such a commonplace cause might be given some grand interpretation!

To reduce the dispersion in the Hubble diagram, HUMASON, MAYALL and SANDAGE concentrated their study on cluster galaxies. This allowed them to obtain a mean velocity and a mean magnitude by measuring several galaxies per cluster. This method also excludes small galaxies. The slope obtained is the same, except that the dispersion is only 0.3 mag., showing that there is no irregular absorption and that the expansion is isotropic.

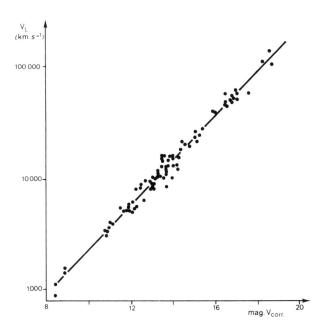

Fig. 3.1. The Hubble diagram for the brightest galaxies in 83 clusters

Subsequently, SANDAGE and his collaborators have measured photoelectrically the magnitudes of the brightest galaxies in 83 clusters. BAUM has extended the possibilities for measuring radial velocities not by measuring the shifts of lines, which are often difficult to see at great distances, but by measuring the general shift of the spectral distribution by photometric means, in several wavelength bands, thus eliminating the necessity for the K-correction. Figure 3.1 collates the various measurements which still lead to a linear relation of slope 0.2 over a velocity range between 1000 and 150,000 km s^{-1}. A slope of 0.2 has also been obtained by CORWIN in a study of ABELL's clusters up to speeds of 60,000 km s^{-1}.

When investigating more closely this diagram, RUBIN showed that, for radial velocities in the range 3000 - 10,000 km s^{-1}, it is possible to detect the motion of the Local Group of 450 km s^{-1} mentioned earlier; however, it is undetectable for higher velocities.

The expansion of the universe is a linear function of distance, distance here being only relative, and measured by the apparent luminosity of the brightest galaxy in a cluster. The law is valid out to a hundred times the distance of the *Virgo* cluster; using classical arguments and anticipating the calibration of the distances, this distance is more than a gigaparsec. Light takes three billion years to reach us from such a distance. The linearity obtained shows that, apart from a rather improbable balance effect, the intrinsic luminosities of the brightest galaxies in clusters have not evolved in any appreciable way over that interval of time. The theoretical work of TINSLEY, furthermore, indicates that the absolute magnitude of giant elliptical galaxies has probably not varied by more than 0.2 mag. in the last three billion years. Further out it is necessary to take account of relativistic effects, as we shall see in Chapter 14.

3.2 The Value of the Hubble Constant

Even if the linearity of the law of expansion is now well established to spectral shifts $\Delta\lambda/\lambda$ around 0.5, the same cannot be said about the value of the constant of proportionality H in equation (3.2). The difficulty results from the fact that to determine H, it is necessary to use a group or cluster of galaxies that is quite close so that its distance in centimeters can be found from absolute distance indicators as described in Chapter 1, yet it has to be far enough away for the recession velocity to be much greater than any peculiar velocity it has. A group or cluster is preferable to an isolated galaxy as it allows more accurate averages to be obtained; it would be even more preferable to make use of several groups or clusters. A good approach is to apply a long range distance indicator, such as the diameter-luminosity relation, to a dozen or so of ABELL's nearest clusters, which have radial velocities of order 10,000 km s^{-1}.

In fact, the actual state of the research is such that the *Virgo* cluster is virtually the only suitable candidate. The estimates of the Hubble constant depend on it. But its uniqueness, and a recession speed which certainly does not exceed by an order of magnitude the individual velocities of the galaxies or possible overall motions in the Local Supercluster, explain why H can still undergo revisions reaching 50%.

One of the most careful studies of the *Virgo* cluster has been done by HOLMBERG. The main difficulty lies in excluding the non-member galaxies which are situated behind and in front of the cluster when calculating the average recession speed of the cluster. Evaluating the distances by a method based on the relationship he found between the absolute magnitude, colour index and mean surface brightness, HOLMBERG obtained

$$\mu = 30.5 \pm 0.9 \tag{3.5}$$

as the mean distance modulus of the cluster.

The mean radial velocity V_L, with respect to the Local Group is

$$V_L = (1016 \pm 60) \text{ km s}^{-1} \tag{3.6}$$

which together with (3.5) leads to

$$H = (80 \pm 35) \text{ km s}^{-1} \text{ Mpc}^{-1}. \tag{3.7}$$

By a similar study of the *Leo* group, HOLMBERG obtained $H = 75 \text{ km s}^{-1} \text{ Mpc}^{-1}$.

Applying the diameter-luminosity relation to spiral galaxies belonging to the *Virgo* cluster, calibrated by six standard galaxies, HEIDMANN obtained:

$$\mu = 30.7 \pm 0.4 \tag{3.8}$$

from which, with the velocity (3.6)

$$H = (73 \pm 14) \text{ km s}^{-1} \text{ Mpc}^{-1}. \tag{3.9}$$

TULLY and FISHER, by their rotation velocity-luminosity relation, obtained for the *Virgo* cluster the same value, $\mu = 30.6 \pm 0.2$. Their relation also gives $H = 75 \text{ km s}^{-1} \text{ Mpc}^{-1}$ for the *Ursa Major* group.

SANDAGE and TAMMANN, calibrating VAN DEN BERGH's luminosity classes with their measurements of H II regions, obtained for *Virgo*, also using (3.6):

$$\mu = 31.5 \pm 0.1 \text{ and } H = (52 \pm 6) \text{ km s}^{-1} \text{ Mpc}^{-1}. \tag{3.10}$$

Meanwhile, BOTTINELLI and GOUGUENHEIM have shown that their calibration for bright luminosity classes is too strong and should in fact lead to $H = (76 \pm 8) \text{ km s}^{-1} \text{ Mpc}^{-1}$.

By means of the colour-luminosity relation, VISVANATHAN found for *Virgo* $\mu = 30.6 \pm 0.4$ and thus $H = 75 \text{ km s}^{-1} \text{ Mpc}^{-1}$.

Finally, applying all the available distance criteria to several isolated galaxies, groups or clusters, DE VAUCOULEURS obtained

$$H = (88 \pm 13) \text{ km s}^{-1} \text{ Mpc}^{-1}. \tag{3.11}$$

In what follows, we will adopt for H sometimes 80, sometimes for simplicity 100 km s^{-1} Mpc^{-1}, but an error of some 50 per cent is still possible.

The inverse of H, which is a time, called the *Hubble time*, gives some idea of the age of the universe, for if the expansion has always been at the same rate, the universe would have been a singularity, $H^{-1} = 12.2$ billion years ago, for $H = 80 \text{ km s}^{-1} \text{ Mpc}^{-1}$.

4. Nearby Intergalactic Matter

The universe seems to be populated principally by galaxies; but this appearance results from the fact that, because of their luminosity, they are relatively easy to see. Therefore, if we wish to estimate the density of the universe, only taking into account the galaxies, we may get a lower limit below the true figure. It is thus important to find out what contribution may be attributed to sources other than galaxies, and in particular to gas, dust and other intergalactic objects.

Neutral hydrogen can be studied by means of the Lyman α line or the 21-cm line; ionised hydrogen can be detected by Hα radiation or, in the radio band, by its recombination lines or thermal radiation; dust by the absorption produced; and very hot gas by its X-radiation. All of these studies are very difficult and with the present state of the art it is in general only possible to obtain upper limits of greater or lesser interest. We divide this chapter between optical observations, 21-cm radio observations and X-ray observations.

4.1 Optical Observations

Observations which relate to the presence of dust in intergalactic space are still very sparse. HOLMBERG has measured the colour indices of spirals in the *Virgo* cluster: half are reddened by 0.2 mag., indicating that in places in the cluster clouds of dust exist, producing absorption of around 0.5 mag.

KARACHENTSEV and LIPOVETSKY have analysed the catalogues of clusters by ABELL and by ZWICKY and obtained on average an absorption in the clusters of 0.22 mag. in the blue and 0.13 in the red, which corresponds to a mean density of dust in the universe of $1.6 \cdot 10^{-33}$ g cm^{-3}. Using quasars, one is even able to lower this limit by a factor of 10.

As for dust absorption present in intergalactic space in general, it can be no more than 10^{-4} mag. Mpc^{-1}.

It seems fairly well established that dust exists in clusters of galaxies. However, its contribution to the mass of the cluster is very small, not exceeding one per cent.

In another approach, the existence of intergalactic material can be deduced, if not the amount, by the observation of peculiar galactic structures. Thus, ZWICKY,

ARP and VORONTSOV-VEL'YAMINOV have given many an example of galaxies with jets, appendages or bridges extending out to several galactic diameters indicating that material, which is probably gaseous and/or stellar, may be dispersing into intergalactic space.

4.2 21-cm Radio Observations

Neutral hydrogen is detectable by its 21-cm line, due to the hyperfine structure of its ground state.

Observations in emission - In emission, the situation is, theoretically, the following: when a region of sky is observed in the beam of a radio telescope, the beam width of the antenna is determined by diffraction and has angular size of the order λ/ϕ where λ is the wavelength and ϕ the diameter of the radio telescope. The observed brightness is given by the curve in Fig. 4.1: there is a general *background continuum* CC' due to thermal and synchrotron emission from various sources: the galactic halo, distant radio sources. Around 21-cm in wavelength, an excess G is produced by neutral hydrogen in our Galaxy. Then an excess emission IG begins, due to neutral hydrogen distributed uniformly in intergalactic space and observed at progressively longer and longer wavelengths, in proportion as one considers the atoms further and further away. Because of the expansion of the universe, their 21-cm lines suffer positive spectral shifts. Finally, if there is a cluster of galaxies in the observed region, which contains more intergalactic hydrogen, an excess of radiation A will be observed with a spectral shift $\Delta\lambda$, corresponding to its radial velocity.

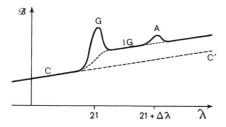

Fig. 4.1. A radio spectrum in the region of the 21-cm wavelength

In practical radio astronomy, wavelength is replaced by radial velocity v according to the Doppler relation

$$v = c\Delta\lambda/\lambda \tag{4.1}$$

and the brightness $\mathscr{B}$ by the *brightness temperature* T_b according to the Rayleigh relation

$$T_b = \mathscr{B}\lambda^2/2k \tag{4.2}$$

where k is the Boltzmann constant. The data from 21-cm line observations are then *line profiles*, which give $T_b(v)$ when the background continuum CC' has been subtracted.

The case of a cluster - The profile A, integrated in velocity, gives the area of the line due to neutral hydrogen atoms contained in the cluster and seen in the beam of the radio telescope. This *integrated profile* I_b is proportional to the average thickness of the cluster l and the mean concentration of atoms in the cluster n:

$$I_b \equiv \int_A T_b(v) \, dv = \beta \ln \tag{4.3}$$

where β is an atomic constant, which may be calculated from the transition probability for the two hyperfine states: $\beta = 5.4865 \cdot 10^{-14}$ degree cm^3 s^{-1}. The formula assumes that the atoms in the cluster do not conceal one another, that is to say, that the optical thickness in the 21-cm line is small, as is the general case.

By directing the beam of the radio telescope at various points over the cluster, denoted by rectangular coordinates (x, y) and integrating over the sky the integrated profiles I_b, the total flux of radiation in the 21-cm line emitted by the whole cluster may be calculated:

$$F_H = 1.24 \cdot 10^{-3} \int T_b(v, x, y) \, dv \, dx \, dy; \tag{4.4}$$

for T_b in degrees Kelvin, v in km s^{-1} and x and y in minutes of arc, F_H is expressed in 10^6 solar masses Mpc^{-2}; the mass of neutral hydrogen which it contains is then simply, in 10^6 $M_\odot$

$$\mathscr{M}_H = F_H D^2, \tag{4.5}$$

D being the distance in Mpc.

ROBERTS and others have observed in this way the principal clusters, and only obtained upper limits of the order of 10^{11} $M_\odot$. By contrast, several isolated clouds of neutral hydrogen have been found in the *Sculptor* group, for example, with masses of approximately 10^8 $M_\odot$. But these isolated clouds have masses less than $6 \cdot 10^9$ $M_\odot$, and there are fewer of them than 0.1 Mpc^{-3}.

The case of intergalactic space - The emission from hydrogen atoms distributed uniformly in space manifests itself as a rise IG above the continuum CC' in Fig. 4.1. Because of the expansion of the universe, an interval of velocity dv corresponds to an interval of distance dl = dv/H. Following formula (4.3) the rise T_b is such that

$$dI_b = T_b \, dv = \beta(dv/H)n$$

where

$$n = 5 \cdot 10^{-5} T_b \tag{4.6}$$

for $H = 80$ km s^{-1} Mpc^{-1}, T_b in degrees and n in number of atoms cm^{-3}.

PENZIAS and SCOTT obtained only an upper limit for this rise, leading to

$$n < 4 \cdot 10^{-6} \text{ cm}^{-3}$$

or an upper limit in density of $7 \cdot 10^{-30}$ g cm^{-3}.

Observations in absorption - Case of a cluster - Imagine observing with a radio telescope R a radio source S subtending a solid angle Ω_s, small in comparison with the beam of the antenna Ω_a and in front of which there is a cloud of atomic hydrogen H, larger than Ω_a (Fig. 4.2). The telescope receives two types of radiation: 1 and 2.

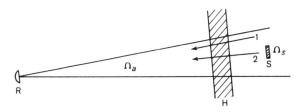

Fig. 4.2. An observation in absorption

Outside the 21-cm line, radiation 1 has a brightness temperature $T_1 = 0$, and radiation 2 a temperature $T_2 = T_c$ where T_c is the brightness temperature of the source in the continuum. The temperature observed at the antenna is then proportional to T_c and Ω_s:

$$T_a' = g_0 \, \Omega_s \, T_c \tag{4.7}$$

where g_0 characterises the antenna gain.

In the line, on the other hand, the cloud radiates and has a brightness temperature T_1 depending on its optical depth τ and the way in which the two hyperfine levels are populated. Using the Boltzmann law, the population of each level i may be represented by $\exp(-E_i/kT_s)$ where E_i is the energy of the level and T_s is a quantity called the *spin temperature*. One has then for radiation 1

$$T_1 = T_s(1 - e^{-\tau}). \tag{4.8}$$

For radiation 2, there is the same emission together with the absorbed emission of the radio source

$$T_2 = T_s(1 - e^{-\tau}) + T_c \, e^{-\tau}.$$

The antenna temperature in the line is then

$$T_a'' = g_0 \, \Omega_s \, T_2 + g_0 \, \Omega_a \, T_1. \tag{4.9}$$

The quantity $g_0 \Omega_a$ is the *beam efficiency* ρ of the antenna. The strength of the line T_a''' with respect to the continuum will be the difference $T_a'' - T_a'$, being, in the usual case when τ is small and $T_s \ll T_c$

$$T_a''' \simeq \tau(\rho T_s - T_a').$$

If, further, the radio source is strong, that is if the antenna temperature which it produces T_a' is large compared with ρT_s, which is usually the case, one simply has

$$T_a''' = -\tau T_a' \tag{4.10}$$

showing that there is an absorption line whose relative depth, T_a'''/T_a' is given directly by the optical depth τ.

Atomic theory shows that the absorption profile integrated over the radial velocities for which the line is observed, is

$$\int \tau(v) \, dv = \beta \ln/T_s \tag{4.11}$$

where l and n are the mean thickness and concentration of the absorbing cloud and β the constant we have already met. The term in $1/T_s$ comes from the fact that, if T_s is large, the populations of the two hyperfine levels equalise and absorption diminishes.

Finally, the characteristics of the hydrogen cloud are given by

$$\ln = 5.9 \cdot 10^{-7} T_s \int \tau(v) \, dv \tag{4.12}$$

where l is in Mpc, n in atoms cm^{-3}, T_s in degrees and v in km s^{-1}.

Unfortunately, T_s comes into this equation. This spin temperature, which governs the relative populations of the two hyperfine levels in intergalactic space, is very difficult to evaluate, but is thought to be in the region of 3 K.

The *Virgo* cluster of galaxies contains a powerful radio source, NGC 4486, and a powerful quasar, 3C 273, is lying behind it. ALLEN, in observing these two sources, has found $\tau < 10^{-3}$. With $T_s = 3°$, l = 3 Mpc and assuming that the hydrogen has the same velocity dispersion as the galaxies, formula (4.12) gives $n < 7 \cdot 10^{-7}$ cm^{-3}, a density less than $1 \cdot 10^{-30}$ g cm^{-3}. Supposing the neutral hydrogen being spherically distributed throughout the cluster, its mass is $\mathcal{M}_H < 10^{12}$ $M_\odot$, an upper limit of greater interest than that obtained by emission measurements.

Some less precise but more numerous measurements have been made on about forty more distant clusters, leading to $\tau < 0.1$ and implying, statistically, that even in the form of intergalactic clouds, neutral hydrogen is not an important contributor to the mass of clusters. Several cases of 21-cm absorption were observed, but they are due to hydrogen associated with galaxies lying in the line of sight.

The case of intergalactic space - If intergalactic space contains neutral hydrogen which is uniformly distributed and participates in the expansion of the universe, the continuum spectrum of a radio source will show a dip IG between radial velocities

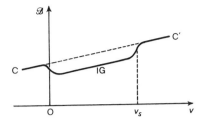

Fig. 4.3. Absorption due to intergalactic neutral hydrogen

zero and that of the source v_s (Fig. 4.3); this dip is due to the absorption of continuum radiation by the atoms situated between us and the source. An interval of distance dl corresponds to an interval of velocity dv = H dl, and equation (4.11) gives

$$\tau(v) \, dv = \beta \, dl \, n/T_s$$

or

$$n = \tau H T_s/\beta = 3 \cdot 10^{-5} \, \tau T_s \qquad (4.13)$$

where n is the concentration of neutral hydrogen atoms cm^{-3}, τ is the optical depth, T_s the spin temperature in degrees and H = 80 km s^{-1} Mpc^{-1}.

Observations of the strongest source of radio continuum emission, *Cygnus A*, which has a recession speed of 17,000 km s^{-1}, have so far given negative results. PENZIAS and SCOTT obtained n < 7 · 10^{-8} cm^{-3} or a density less than 1 · 10^{-31} g cm^{-3}, a hundred times weaker than the density ρ_{EdS} of the Einstein-de Sitter model which we shall study in Chapter 10. This method is much more sensitive than that using measurements in emission.

PEEBLES has studied the effect it would have on the observations if, instead of a uniform distribution, the hydrogen were in clouds; the atoms would then be able to conceal each other partially, and PEEBLES found a density less than 0.3 ρ_{EdS}.

The direction of *Cygnus A* is not an isolated case. Towards the source *Fornax A*, with radial velocity 1600 km s^{-1}, though a little less precise, the observations reveal no dip.

In the fourth part of this book which is devoted to observations in the relativistic zone, we will present the information about neutral hydrogen which the study of quasars provides by means of the Lyman α line, and about the hydrogen molecule through the Lyman band. There we will also tackle the question of the possible existence of ionized hydrogen distributed through intergalactic space in general.

4.3 X-Ray Observations

X-ray detectors sensitive to energies between 2 and 40 keV, on satellites, have shown that about thirty of ABELL's clusters are powerful X-ray sources. Their spectra are compatible with that produced by an ionized gas - hydrogen and helium - at a very high temperature, by bremsstrahlung; that is, by the emission of X-ray photons when electrons are deflected as they pass through the Coulomb fields of nuclei.

Observation of the *Perseus* cluster has shown that a quarter of the emission is coming from the active central galaxy NGC 1275, the rest coming from a diffuse region more than 1 Mpc in diameter. These vast emission halos have been found by GIACCONI in several clusters.

Typically, these clusters emit 10^{45} erg s^{-1} between 2 and 10 keV. The gas in the clusters is at 10^8 K, has a mass of 10^{15} $M_\odot$ and a radius of approximately 10 Mpc.

GIACCONI further discovered superclusters containing around six clusters, a hundred times more powerful still. Very probably, the emission is produced by bremsstrahlung in a mass of 10^{16} $M_\odot$ of ionized gas contained in a volume 50 Mpc in diameter. This mass is ten times greater than that of the associated galaxies and is almost sufficient to bind gravitationally the member clusters of these superclusters. Lines of highly ionized iron have been detected around 7 keV; they imply that the gas is not of primordial origin, but has undergone stellar nucleosynthesis; it has probably been torn out of the halos of galaxies during close interactions.

The existence of hot gas in clusters is confirmed by the diffusion it produces through the Compton effect, on the cosmic background at 3 K (see Chapter 17); SUNYAEV and ZEL'DOVICH have shown that in crossing such a cluster, these photons would be able to gain energy by interaction with rapid electrons in the gas, and produce a lowering of the temperature of the cosmic background in the Rayleigh-Jeans portion of the Planck spectrum of a black body, given by

$$\Delta T \sim -\frac{2kT}{m_e c^2} \sigma_T \int n\, T\, dl \qquad (4.14)$$

where σ_T is the Thompson cross-section (see page 141), m_e and T the mass and kinetic temperature of the electrons. This prediction has been confirmed by LAKE and PARTRIDGE who, observing clusters at 9 mm wavelength, have found $\Delta T = 1$ to $3 \cdot 10^{-3}$ K and $n \sim 10$ cm^{-3}.

Note that another fact allowed the existence of gas in clusters to be foreseen: the discovery of tail radio galaxies and their interpretation by the relative movement of these objects with respect to an intergalactic medium.

5. The Density of the Universe

Having collected all the available information on the distribution of galaxies in space and on intergalactic material, we are able to attack the problem of the density of the universe. This density is one of the four fundamental parameters observable within a gigaparsec and contributing to cosmology. The other three are the distance scale, the Hubble constant which we have already studied, and the age of the universe which is the subject of the next chapter.

5.1 The Contribution of Galaxies

In order to evaluate the contribution which galaxies make to the density of the universe, we are going to begin by reviewing the data obtained with the successive steps in the exploration of space.

1. *The Local Group of Galaxies* - Its mean density is $2.15 \cdot 10^{11}$ $M_\odot$ Mpc^{-3} or $1.5 \cdot 10^{-29}$ g cm^{-3}.

2. *Groups of Galaxies within 10 Mpc* - A group equivalent to the Local Group every 7 Mpc leads to $2 \cdot 10^9$ $M_\odot$ Mpc^{-3} or $1.5 \cdot 10^{-31}$ g cm^{-3}. The enormous drop in mean density in passing from the Local Group to a region of dimensions ten times greater is worthy of note.

3. *Coma Cluster* - Even if the distribution of light over the *Coma* cluster $E(r)$ is relatively well known (Section 2.4), the distribution of mass, which is what interests us in obtaining the density of the universe, is very poorly known. If one supposes that the ratio $\mathcal{M}/L_{pg}$ of the mass to the luminosity is, in the *Coma* cluster, independent of r, the density $\rho(r) = (\mathcal{M}/L_{pg}) E(r)$ will vary as $1/r^2$ like $E(r)$.

This variation is the same as in a self-gravitating isothermal sphere, which means that $(\mathcal{M}/L_{pg})$ may be calculated. Let us consider in effect a spherical distribution of particles held under their own gravitation. In equilibrium, there is a balance, at every distance r from the centre, between the gradient of the gas pressure p and the force of gravity per unit volume:

$$dp/dr = -\frac{G\rho}{r^2} \int_0^r 4\pi r^2 \rho \, dr$$

where ρ is the density and G the constant of gravitation.

Taking the derivative of this relation with respect to r, and replacing the pressure p by the equation of state of a gas made up of particles of mass m at a temperature T

$$p = \frac{\rho}{m} kT \tag{5.1}$$

one obtains a differential equation giving $\rho(r)$:

$$\frac{d}{dr}\left(\frac{r^2}{\rho}\frac{d\rho}{dr}\right) + \frac{4\pi Gm}{kT}\rho r^2 = 0. \tag{5.2}$$

This is Emden's equation, of which the particular solution is

$$\rho(r) = (kT/2\pi Gm)/r^2. \tag{5.3}$$

Comparing with the density of the *Coma* cluster, one obtains

$$(\mathcal{M}/L_{pg}) = \pi kT/2Gc_o \, m,$$

with the notation of Chapter 2, Section 4.

Now, if v is the space velocity of the particles, one has on average $m<v^2>/2 = 3kT/2$, and also $<v^2> = 3\sigma^2$, where σ is the radial velocity dispersion (radial as observed from the Earth).
Thus

$$(\mathcal{M}/L_{pg}) = \pi\sigma^2/2Gc_o. \tag{5.4}$$

With the numerical values of Chapter 2, (5.4) gives $(\mathcal{M}/L_{pg}) = 120$ in solar units. The luminosity obtained within $r = 100$ arc min - being 2.7 Mpc - leads then to a mass $\mathcal{M} = 7.6 \cdot 10^{14} \, M_\odot$ and the corresponding mean density of the cluster is $1.2 \cdot 10^{13} \, M_\odot$ Mpc^{-3}, or $8 \cdot 10^{-28}$ g cm^{-3}.

Two remarks must be made: this evaluation rests on the use of a theory which assumes that the *Coma* cluster is in a stationary state. Though effectively, the cluster shows certain characteristics of relaxation, it should not be forgotten that interactions between galaxies are very rare; it already takes three billion years for a galaxy to pass right through the cluster. This relaxation, if it really exists, must date from before the condensation of protogalactic clouds, when, because they were larger and probably more tightly packed (considering the subsequent expansion of the universe), they were able to exchange momentum and energy. At this epoch, the expansion was rapid, and it would be necessary to take account of this in the theory.

Second remark: the ratio $(\mathcal{M}/L_{pg})$ is large when compared with those of galaxies. For elliptical galaxies - major constituents of the cluster - HOLMBERG obtained $(\mathcal{M}/L_{pg}) = 50$. On the other hand, the ionized gas revealed by its X-radiation may give a ratio similar to the cluster value.

4. *Abell's Clusters* - A rich cluster every 175,000 Mpc^3; assuming that they are equivalent to the *Coma* cluster, they produce a mean density of $4.3 \cdot 10^9 \, M_\odot$ Mpc^{-3} or $3 \cdot 10^{-31}$ g cm^{-3}. This is a lower limit on the density of the universe since the rich clusters are not its only constituents.

5. *Galaxy counts* - The counts of galaxies as a function of their apparent magnitude, over vast areas of sky, also give an estimate of the density of the universe.

The various luminosity functions of galaxies in Fig. 2.1 all have almost the same shape, one being obtainable from another by a simple vertical shift. In a general way, the number dn of galaxies in an interval of absolute magnitude dM per Mpc^3, may then be written

$$dn = C\Phi(M) \, dM \tag{5.5}$$

where Φ gives the form and C the numerical value. The number of galaxies observable in the interval dM, down to apparent magnitude m, for uniformly populated space, is

$$\frac{4\pi}{3} 10^{0.6(m-M-25)} C\Phi(M) \, dM$$

where $10^{0.6(m-M-25)}$ is simply the cube of the distance corresponding to m. Integrating over M, one obtains the total number of galaxies observable down to magnitude m:

$$N(m) = \frac{4\pi}{3} 10^{-15} 10^{0.6m} C \int_{-\infty}^{\infty} 10^{-0.6M} \Phi(M) \, dM. \tag{5.6}$$

$N(m)$ is given by the surveys, equation (2.2), and $\Phi(M)$ by the curves in Fig. 2.1, which allows the value of C to be deduced. One may then calculate the luminosity per Mpc^3

$$L = \int_{-\infty}^{\infty} C\Phi(M) \, 10^{-0.4M} \, dM \tag{5.7}$$

where $10^{-0.4M}$ is the luminosity corresponding to absolute magnitude M.

By this method, OORT and SHANE have found

$$L_{pg} = 0.2 \cdot 10^9 \, L_\odot \, Mpc^{-3}.$$

In order to convert to density, it is necessary to use the ratios $(\mathcal{M}/L_{pg})$ for galaxies. According to HOLMBERG, this ratio has the value 50, in solar units, for elliptical galaxies, and according to N. HEIDMANN and ROBERTS, it is 5 for spiral and irregular galaxies. Ellipticals account for 20% of galaxies. The mean value of $(\mathcal{M}/L_{pg})$ is thus of the order of 14, which leads to a mean density of $3 \cdot 10^9 \, M_\odot \, Mpc^{-3}$ or $2 \cdot 10^{-31} \, g \, cm^{-3}$.

5.2 The Contribution of Intergalactic Material

Let us collect together the data of Chapter 4 and, likewise, that which we will present in Chapter 15.

In clusters of galaxies, dust contributes at most 1% of the mass of the galaxies. For the clusters, the masses of neutral hydrogen are less than $10^{11} \, M_\odot$. With a mean ratio $(\mathcal{M}/L_{pg}) \approx 20$ for the *Virgo* cluster, about 60% of its population being spirals, the total mass is $\approx 2.4 \cdot 10^{13} \, M_\odot$; neutral hydrogen contributes, then, less than 0.4%.

For the *Coma* cluster, with the mass given in the preceding section, this contribution is less than 0.01%. It may be said, then, that the contribution of neutral hydrogen in clusters is negligible.

In intergalactic space in general, the measurement of the 21-cm line in absorption leads to a density of neutral hydrogen less than 10^{-31} g cm^{-3}. We will see in Chapter 15 that measurements on quasars lead to a density of neutral hydrogen less than $5 \cdot 10^{-35}$ g cm^{-3} and to a density of molecular hydrogen less than 10^{-32} g cm^{-3}. However, it should be said that these measurements, made on objects thought to be very distant, correspond to the very distant past of the universe, and that perhaps, at that epoch, the densities of these gases were much smaller than now if the hydrogen was at that time highly ionized.

The ionized hydrogen within clusters, discovered in large clusters by its X-rays, makes a contribution ten times greater than their galaxies. It may then have a mean density of $3 \cdot 10^{-30}$ g cm^{-3}.

5.3 Other Contributions

Another contribution to the density of the universe comes from the energy of the radiation which pervades it, since, according to Einstein's relation, an energy E is equivalent to a mass E/c^2.

In the electromagnetic spectrum, from X-rays to radio waves, the most important part comes from the cosmic background radiation at 3 K, observed between wavelengths of 1 mm and 1 m. Its total energy corresponds only to a density of $5 \cdot 10^{-34}$ g cm^{-3}.

Among the rays consisting of particles, cosmic rays probably play the most significant role: 1 eV per cm^3, or 10^{-33} g cm^{-3}. But according to NOVIKOV and ZEL'DOVICH, neutrinos may have a density comparable with the 3 K cosmic background.

Do any other contributors to the density of the universe exist? The question of ionized hydrogen between clusters has still not been completely resolved. There are practically no direct indications from the observations that lead to a general background of intergalactic ionized hydrogen with a density as great as ρ_{EdS}, or 10^{-29} g cm^{-3} (see Chapter 10). But, nevertheless, its density could be distinctly greater than the general density due to the galaxies, which is several times 10^{-31} g cm^{-3}.

Fairly strong arguments have been presented in favour of "hidden mass" in the universe: study of our own Galaxy in the neighbourhood of the Sun, study of groups and clusters of galaxies via the virial theorem, the possible existence of black holes.

Galaxies may be surrounded by large, massive halos; several facts suggest this: photography on very sensitive plates, X-ray emission surrounding giant ellipticals, rotation velocities which do not decrease as one moves away from the central regions of certain spirals. Most of the mass of these galaxies could reside in their halos where the ratio $\mathscr{M}/L$ might reach 1000 according to SPINRAD and where the population could probably consist of very low mass stars. For some galaxies, at least, their masses may be ten times greater than the "classical" masses attributed to them before about 1977.

5.4 The Density of the Universe

Table 5.1 collects together the various items of information on density from the preceding paragraphs, with relevant comments on how they relate to the density of the universe.

Table 5.1.

Source	Density (g cm^{-3})	Remarks
Local Group	$1.5 \cdot 10^{-29}$	too large
groups of galaxies	$1.5 \cdot 10^{-31}$	lower limit
Coma cluster	$8 \cdot 10^{-28}$	much too large
Abell clusters	$3 \cdot 10^{-31}$	lower limit
counts of galaxies	$2 \cdot 10^{-31}$	lower limit
neutral hydrogen	$< 10^{-31}$ or $5 \cdot 10^{-35}$	negligible
molecular hydrogen	$< 10^{-32}$	negligible in the past
ionized hydrogen	$3 \cdot 10^{-30}$?	the most important ?
photons	$5 \cdot 10^{-34}$	negligible
cosmic radiation	10^{-33}	negligible
neutrinos	$\sim 10^{-33}$?	negligible
black holes	$\lesssim 10^{-30}$?	not known

Without taking into account neither possible hidden mass - ionized hydrogen between clusters or black holes - nor X-ray observations, the mean density of the observable universe is several times 10^{-31} g cm^{-3}, the value

$$\rho_0 = 3 \cdot 10^{-31} \text{ g cm}^{-3} \tag{5.8}$$

appearing to be a reasonable estimate. But if the X-ray observations of large clusters are confirmed, the density of the universe may reach the value

$$\rho_0 = 3 \cdot 10^{-30} \text{ g cm}^{-3}. \tag{5.9}$$

On the basis of a density-radius relationship for astronomical objects, DE VAUCOULEURS suggests that, if observations penetrated further into space, the mean density may decrease, if the material in the universe were distributed in clusters of order n which are themselves grouped in clusters of order n + 1, following a hierarchial model proposed at the beginning of the century by CHARLIER and studied by WERTZ for the case of an expanding universe.

In conclusion, the observations to within a gigaparsec lead to the value given by (5.8) or perhaps (5.9) for the mean density of the universe. But one must not lose sight of the fact that this density might be greater if hidden mass exists, or smaller if the universe is hierarchical.

6. The Age of the Universe

The best lower limit for the age of the universe is provided by a combination of theory and observation, by the theory of stellar evolution and the observation of globular star clusters.

Around 200 globular clusters, each containing 10^5 to 10^6 stars, are in orbit around our Galaxy, forming a halo with a radius close to 20 kpc. After twenty years' work, interpreting precise photoelectric measurements down to 22nd magnitude, made with the 5-m telescope at Mount Palomar, SANDAGE has obtained U, B and V magnitudes for stars in four globular clusters, and has constructed the corresponding colour-magnitude diagrams.

The distribution of points on such diagrams can be understood in terms of stars born at the same time, with the same initial chemical compositions, but different masses. The chemical composition is specified by X, Y and Z, the relative proportions by mass of hydrogen, helium and heavier elements. Thus, for X = 0.7, Y = 0.3 and Z = 0.001, after a time $T = 12.5 \cdot 10^9$ years, the stars of initial mass $\mathcal{M}$ less than 0.7 $M_\odot$ are on the main sequence; for $\mathcal{M} = 0.75\ M_\odot$ they are at the turn-off point and for $\mathcal{M}$ between 0.75 and 0.78 $M_\odot$ they are on the red giant branch; for $\mathcal{M} \simeq 0.8\ M_\odot$ the stars may, in the course of their lives, throw out up to a quarter of their mass; they are, on the whole, situated on the horizontal branch, but also on the asymptotic branch. Finally, for any greater initial mass, the stars will have become white dwarfs.

This information results from stellar models in which temperature, pressure, convection, diffusion and thermonuclear reactions are evaluated at various depths in the star. These calculations give the paths which stars follow in the colour-magnitude diagram according to the values of X, Y and Z; the positions of the points at a time equal to the age of a cluster T should reproduce the observed diagram.

Z is given by spectroscopic observations and has a value between 10^{-3} and 10^{-4}. The models calculated by IBEN and ROOD, and by DEMARQUE and his collaborators allow Y and T to be estimated, mainly from the main sequence turn-off point.

The values of T for the various clusters range from $1 \cdot 10^{10}$ to $1.4 \cdot 10^{10}$ years, with an error of $3 \cdot 10^9$ years. One of the most important results is that the ages of the different clusters all agree, with the same value

$$T = (12 \pm 3) \cdot 10^9 \text{ years}. \tag{6.1}$$

A second remarkable result is that the initial proportion of helium is found to be

$$Y = (29 \pm 3) \%, \tag{6.2}$$

practically equal to that found in the cosmological interpretation of the 3 K cosmic background. In this interpretation, as we shall see in Chapter 17, the first quarter of an hour following the start of the expansion of the universe from its condensed state, was occupied by an active series of nuclear reactions leading to the transformation of a quarter of the material into helium.

In the same chapter we will see that our own Galaxy must have formed very rapidly after the start of the expansion; one might think that only 100 or 200 million years must have elapsed between the beginning of the universe and the formation of the globular star clusters. The age of the universe may, then, be put equal to the value in (6.1), $(12 \pm 3) \cdot 10^9$ years.

Part II
Spaces with Constant Curvature

Spaces with constant curvature are the foundation of the simplest cosmological models of the universe: the uniform models of General Relativity. At the outset these models provide us with a mathematical base, since without them it is impossible to establish rigorously the exact expressions that replace the inexact formulas of classical geometry and mechanics. Furthermore, they describe the physical substratum, to a first approximation at least, of the real space in which we live. In fact, our everyday life does not require the concept of curved space; however, there is very little probability - and there is not even any reason - for the space which we experience to have classical Euclidean geometry.

If theodolites had a much greater precision than they in fact have, then we would perhaps have seen that the sum of the angles in a triangular field would differ from $180°$. In that case the concept of curved space would be more familiar. Everyone would know, from experience, that if one moved in a straight line in spherical space one would eventually return from the opposite direction to the starting point without having deviated in any manner whatever from a straight line.

In fact, since the publication of the General Theory of Relativity more than half a century ago, the notion of curved space has gained ground. The picture that we generally have of curved space is given by the surface of a sphere. On this a "straight" line - the shortest route between two points - is an arc of a great circle. However, although this model is easy to visualise, it is only valid for two-dimensional spherical space. To extend to three-dimensional spherical space we have to adopt the three-dimensional surface of a hypersphere embedded in four-dimensional Euclidean space, which is hardly easy to portray.

The simple model of two-dimensional spherical space, provided by the surface of a sphere, is of little help and brings the risk of our making serious errors in the three-dimensional case. Furthermore, there is no direct equivalent of two-dimensional hyperbolic space: the sphere would be imaginary and, once again, difficult to visualise. Very often, one falls back on a mental picture of the shape of a mountain pass or saddle, but this representation is false in many respects.

So, in order to describe the curvature of space, we must use mathematics. But, without giving way to misleading pictures, we will avoid a purely mathematical presentation. Instead, we shall often illustrate the argument with physical aspects which will give to our investigation of curved space a similarity to a real exploration of the universe in which we live. We will be paying special attention to spaces with constant curvature, because these, after all, are the simplest and most frequently used cases. Finally, we shall describe several exact models which help us to visualise and become more familiar with the properties of such spaces.

In this second part of the book, which deals with spaces of constant curvature, there are three chapters. In the first one, starting from classical Euclidean geometry, we give a brief summary of the essentials of tensor calculus, after which we treat spaces that are locally Euclidean. By this we mean that they have no curvature, but the topological structure is not necessarily that of classical Euclidean space.

The second chapter is concerned with spaces that are locally non-Euclidean, with constant curvature not necessarily zero at every point. If the curvature is positive then the space is locally spherical, and if it is negative the space is locally hyperbolic. The special case of zero curvature brings us back to locally Euclidean space.

At this point it is worth considering the unfortunate terminology by which Euclidean space is to be regarded as a particular type of non-Euclidean space. The term "non-Euclidean" is enshrined in history and, in fact, for us it signifies "constant curvature". In the case of surfaces, the corresponding nomenclature is "non-planar" for a curved surface, and a plane surface would then be a special case of a "non-planar" surface. To remind ourselves of this historical anomaly we shall always hyphenate the expression "non-Euclidean".

Finally, in the third chapter, we describe the special cases of simply-connected non-Euclidean spaces: spherical space with positive curvature, and hyperbolic space with negative curvature (and Euclidean space of zero curvature).

This section of the book is mainly inspired by E. CARTAN's remarkable treatise on the Geometry of Riemannian Spaces.

7. Locally Euclidean Spaces

We shall consider a classical Euclidean space in which the points $\underline{M}$ are marked by a system of general coordinates y^i where $i = 1,2,\ldots n$, n being the number of dimensions. This can be any number, but for definiteness we may think of it as equal to 3. In this space a point $\underline{M}'$, infinitesimally close to $\underline{M}$, has coordinates differing from those of $\underline{M}$ by dy^i and is separated from $\underline{M}$ by a distance $ds = |\underline{M}' - \underline{M}|$, given by the *metric element*

$$ds^2 = g_{ij} \, dy^i \, dy^j \quad ^{1} \tag{7.1}$$

where the set of functions g_{ij} are functions of the coordinates y^k. The functions $g_{ij} = g_{ji}$ make a symmetric tensor, the *metric tensor*. As we shall see, it is this tensor that determines the type of coordinates used and the space in which they are embedded.

7.1 Natural Frame

If we vary one of the y^i, then $\underline{M}$ describes a *coordinate line*. n coordinate lines pass through $\underline{M}$. Let $\underline{e}_i$ be the tangent vectors at $\underline{M}$ to the coordinate lines and their modulus $\sqrt{g_{ii}}$. These vectors are the base vectors of the *natural frame* at $\underline{M}$. This natural frame allows us to specify with respect to $\underline{M}$ infinitesimally neighbouring points $\underline{M}'$ thus

$$\underline{M}' - \underline{M} = d\underline{M} = \underline{e}_i \, dy^i \tag{7.2}$$

where the dy^i are similar to Cartesian coordinates. The shape of the frame, but not its orientation, is completely specified by the scalar products

$$\underline{e}_i \cdot \underline{e}_j = g_{ij}.$$

The first step in exploring the geometry of space is now complete.

The components dy^i are said to be contravariant, in contrast to the components dy_i,

---

[1] In accordance with the standard notation of tensor calculus, whenever an upper and lower index coincide we have to sum over the values of the said index $i = 1,\ldots n$.

indicated by lower indices and obtained thus:

$$dy_i = g_{ij} dy^j,$$

which are said to be covariant. The covariant components dy_i of $d\underline{M}$ are simply the scalar products $\underline{e}_i \cdot d\underline{M}$.

7.2 The Riemann-Christoffel Tensor

Now that we have located points $\underline{M}'$ that neighbour $\underline{M}$ by use of the natural frame at $\underline{M}$, the second step in the exploration of the geometry of space consists of locating the natural frames of neighbouring points $\underline{M}'$. If $\underline{e}'_i$ is a base vector of the natural frame at $\underline{M}'$, we have

$$\underline{e}'_i = \underline{e}_i + \omega_i^j \, \underline{e}_j \tag{7.3}$$

where the ω_i^j are infinitesimal to first order, the contravariant components of $d\underline{e}_i = \underline{e}'_i - \underline{e}_i$, and they are linear functions of the displacement $d\underline{M}'$

$$\omega_i^j = \Gamma_{ik}^j \, dy^k. \tag{7.4}$$

The coefficients Γ_{ik}^j are called *Christoffel symbols of the second type*. They are not tensor components; there are n^3 of them. They are actually functions, which we will derive, of the first-order partial derivatives $\partial_l g_{mn}$[1] of the components of the metric tensor. Knowledge of these coefficients will allow us to locate the natural frame at $\underline{M}'$.

To do this calculation we now introduce *Christoffel symbols of the first type*, defined by lowering an index:

$$\Gamma_{ijk} = g_{jl} \, \Gamma_{ik}^l. \tag{7.5}$$

A first series of relations between the Γ is found by differentiating $\underline{e}_i \cdot \underline{e}_j = g_{ij}$ in order to obtain after rearrangement

$$\Gamma_{ijk} + \Gamma_{ikj} = \partial_i g_{jk}. \tag{7.6}$$

A second series of equations is obtained by considering the two infinitesimal paths:
first made by varying y^i alone, then y^j alone,
second made by varying y^j alone, then y^i alone.

---

[1] The notation used is $\partial_i = \dfrac{\partial}{\partial y^i}$, $\partial_{ij} = \dfrac{\partial^2}{\partial y^i \partial y^j}$.

Now, in Euclidean space, the same point is reached whichever of these paths is chosen. Using equations (7.2), (7.3) and (7.4), and the fact that $\partial_{ij}\underline{M} = \partial_{ji}\underline{M}$, gives $\Gamma_{ik}^{j} = \Gamma_{ki}^{j}$. From (7.5) there is the symmetry relation

$$\Gamma_{ijk} = \Gamma_{kji}. \tag{7.7}$$

Finally, equations (7.6) and (7.7) yield the Christoffel symbols of the first type

$$\Gamma_{ijk} = (\partial_i g_{jk} + \partial_k g_{ij} - \partial_j g_{ik})/2. \tag{7.8}$$

Thus knowledge of the metric tensor g_{ij} and its derivatives allows us to locate the base vectors $\underline{e}_i'$ of the natural frame at a point $\underline{M}'$, infinitesimally close to M, by reference to the natural frame at $\underline{M}$. Explicitly

$$\left. \begin{array}{l} d\underline{e}_i = \Gamma_{ik}^{j} \underline{e}_j \, dy^k \\ \text{or } \partial_k \underline{e}_i = \Gamma_{ik}^{j} \underline{e}_j \end{array} \right\} \tag{7.9}$$

Having completed the second step in exploring the space, we can now express mathematically a fundamental property of Euclidean space, which will lead us to introduce the crucial Riemann-Christoffel tensor.

Let us consider once more the two infinitesimal paths used above. In Euclidean space, whatever path we will follow, we end up not only at the same point, but we obtain the same natural frame. This can be verified by direct calculation, no matter what coordinate system, Cartesian or general, is used. Equations (7.9) allow us to calculate the changes in the vectors $\underline{e}_i$ along the two paths, and noting that these variations are the same, we obtain

$$\partial_{1k}\underline{e}_i = \partial_{kl}\underline{e}_i \quad \text{or} \quad \partial_l(\Gamma_{ik}^{j}\underline{e}_j) = \partial_k(\Gamma_{il}^{j}\underline{e}_j).$$

Making the calculation, we obtain in succession, with the aid of (7.9)

$$\underline{e}_j \partial_l \Gamma_{ik}^{j} + \Gamma_{ik}^{j} \Gamma_{jl}^{h} \underline{e}_h = \underline{e}_j \partial_k \Gamma_{il}^{j} + \Gamma_{il}^{j} \Gamma_{jk}^{h} \underline{e}_h$$

or, changing j for h in those terms where $\underline{e}_j$ occurs,

$$\underline{e}_h(\partial_l \Gamma_{ik}^{h} + \Gamma_{ik}^{j}\Gamma_{jl}^{h}) = \underline{e}_h(\partial_k \Gamma_{il}^{h} + \Gamma_{il}^{j}\Gamma_{jk}^{h}),$$

that is, as $\underline{e}_h \neq 0$,

$$\partial_l \Gamma_{ik}^{h} - \partial_k \Gamma_{il}^{h} + \Gamma_{ik}^{j}\Gamma_{jl}^{h} - \Gamma_{il}^{j}\Gamma_{jk}^{h} = 0. \tag{7.10}$$

One can show that the first member of (7.10) is a four-index tensor R_{ilk}^{h}, antisymmetric with respect to indices l and k, contravariant in respect to h, and covariant in the other three. It is a function of the g_{ij} and their first and second order partial derivatives. This is the *Riemann-Christoffel tensor*.

In Euclidean space the Riemann-Christoffel tensor is null. Thus, in imposing certain relations on the g_{ij}, it turns out that these functions cannot be chosen arbitrarily. If the coordinate system is freely chosen, it must be done in the framework of Euclidean space, a particular framework that imposes certain conditions. This is an important conclusion.

7.3 Locally Euclidean Space

In Euclidean space the Riemann-Christoffel tensor is null. But, is the converse true: does the nullity of this tensor guarantee that the space is Euclidean? The answer is: no. Space in which the Riemann-Christoffel tensor is null is said to be *locally Euclidean*. It is here that we can begin the third stage in the exploration of the geometry of space. This stage will lead us farther in the study of its properties, into the "local" domain.

The essential difference between Euclidean and locally Euclidean spaces is connected with topological considerations and has to do with shape.

Let us take a simple illustration. Consider the geometry of a right circular cylinder in three-dimensional Euclidean space, the geometry being determined by the metric of this space. Thus the "straight lines" in the two-dimensional space that is the cylinder surface are helixes, making extremal the distance from point to point.

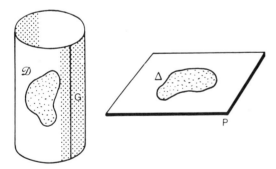

Fig. 7.1. An example of the development of a simply-connected domain

Locally, for any simply-connected domain $\mathcal{D}$ - which can be obtained by tracing a closed curve that does not cut an arbitrarily selected generatrix G - there is an isometric mapping termed the *development* of $\mathcal{D}$ to a simply-connected domain Δ in the Euclidean plane P (Fig. 7.1). For every point in $\mathcal{D}$ there is a corresponding point in Δ, and vice versa. The distance between any two points in $\mathcal{D}$ will be the same as that between the corresponding points of Δ. One example of the mapping is found simply by cutting the cylinder along G and then unwrapping it on the plane P. Δ does not necessarily cover the whole Euclidean plane; at most it covers a strip with parallel sides.

Within 𝒟 all the properties of Euclidean geometry hold: the angles in a triangle sum to 180°; between any pair of points there is one and only one straight line, etc.

However, if we abandon the requirement that 𝒟 is to be simply-connected, then many of these properties cease to hold. Thus, for the domain 𝒟 formed between two circles round the cylinder (Fig. 7.2), there are "straight" routes of finite length: the circles of the cylinder. Through two points there is an infinity of "straight" routes: helixes of various pitches passing through the points. Two straight routes can cut each other many times, for example, if they are helixes of different pitch. The domain 𝒟 under consideration is not simply-connected: a circle round a cylinder cannot be contracted continuously into a point.

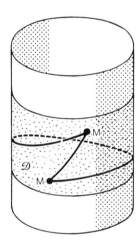

Fig. 7.2. A locally Euclidean domain not simply-connected

Lastly, if we map 𝒟 into a domain Δ on a Euclidean plane, by rolling it on P to impress the "pattern", the mapping is no longer unique. For each point in 𝒟 there is an infinite number of points in Δ. Δ could cover the whole plane, if 𝒟 covered the entire cylinder.

Certain properties of Euclidean space are still valid, for example those concerning the sum of the angles in a triangle. But many of them, as we have seen, are invalid. This example has demonstrated important differences between Euclidean and locally Euclidean spaces.

In fact, the nullness of the Riemann-Christoffel tensor is a pointwise property, expressing the relationships between the values of g_{ij} and their first and second order derivatives, at a point or a set of points. These relations are not sufficiently restrictive to force space to have the particular Euclidean quality, and they leave sufficient freedom by specifying only the more general class of locally Euclidean spaces, whose general properties we shall now study a little more closely.

7.4 Development

The example of the cylinder has given an illustration of *development*. Now consider a point $\underline{M}$ in a locally Euclidean space. The nullness of the Riemann-Christoffel tensor ensures that by making the calculations leading to equations (7.10) in reverse, the position of the natural frame at an infinitesimally nearby point $\underline{M}'$, derivable by the relations (7.9), will be the same whatever way is taken from $\underline{M}$ to $\underline{M}'$, as long as that way is infinitesimally short and does not deviate by some finite distance.

To $\underline{M}$ and its natural frame of vectors $\underline{e}_i$ let us now take, in Euclidean space, a corresponding point $\underline{P}$ and a frame of base vectors $\underline{f}_i$ at $\underline{P}$ with identical form, that is, such that $\underline{f}_i \cdot \underline{f}_j = \underline{e}_i \cdot \underline{e}_j$. To $\underline{M}'$ then, there corresponds $\underline{P}'$ by equation (7.2) and $d\underline{P} = \underline{f}_i dy^i$. Finally, to the natural frame at $\underline{M}'$ we attach the natural frame $\underline{f}'_i$ at $\underline{P}'$ by equations (7.9) and $d\underline{f}_i = \Gamma_i{}^j{}_k \underline{f}_j dy^k$.

These operations give us an isometric mapping of the infinitesimal neighbourhood of $\underline{M}$ in the locally Euclidean space under consideration into the Euclidean space. This mapping procedure is the development.

By integration the development can be extended step by step to a finite neighbourhood of $\underline{M}$. Thus (Fig. 7.3) $\underline{M}$ and the path Γ_1 to $\underline{M}'$ are developed on $\underline{P}$ with the path C_1 going to $\underline{P}'$. And it can be shown that the solution $\underline{P}'$ is unique if the paths connecting $\underline{M}$ to $\underline{M}'$ are compelled to lie in a simply-connected domain $\mathscr{D}$.

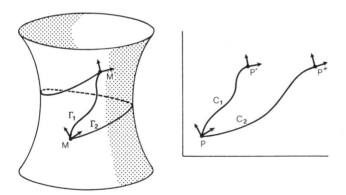

Fig. 7.3. Development of a locally Euclidean space

It is this last restriction that is the nub of the matter. If $\mathscr{D}$ is not simply-connected, then there are paths $\Gamma_2,\ldots,$ from $\underline{M}$ to $\underline{M}'$ such that their developments $C_2,\ldots$ lead to points $\underline{P}''$, different from $\underline{P}'$. Let us take the model of the cylindrical surface, as in the preceding section: note that the various helixes linking M to M' develop according to the straight lines joining P to various points P', P'',....., each with its natural frame $\underline{f}'_i$, $\underline{f}''_i$, all mutually parallel. For an easier understanding of the following paragraphs it is useful to keep the cylinder example in mind.

7.5 Holonomy Groups

Let us take in particular as point M' the point M itself, i.e. consider the development of paths leaving M and arriving at M. We shall thus obtain for the development of M and its natural frame $\underline{e}_i$ the points $\underline{P},\underline{Q},\underline{R}$,..., with their natural frames $\underline{f}_i$, $\underline{g}_i$, $\underline{h}_i$,..., and only P and $\underline{f}_i$ alone if the paths are in a simply-connected domain $\mathscr{D}$ (Fig. 7.4). The displacements D_Q, D_R,... that transform, in Euclidean space, P and $\underline{f}_i$ into other points Q, R, (the so-called homologous points for P) and their frames $\underline{g}_i$, $\underline{h}_i$, form a group. This group is called the *holonomy group* at M of the locally Euclidean space under consideration.

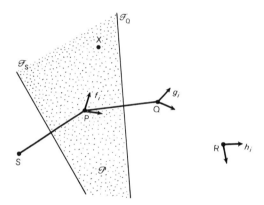

Fig. 7.4. Holonomy group and fundamental polyhedron

This group is generated by *generating operations*; these accomplish, by multiplication, the reproduction of all the displacements D_Q... For the cylinder these generating operations are the identity transformation (1) and the translation (T) passing from P to the nearest homologous point Q.

It can be shown that

1) The group is *discontinuous*; the homologous points Q, R,..., of the point P are separated from P by an amount greater than or equal to a "critical" distance, which for the cylinder is equal to the circumference of the cylinder;

2) Generating operations, with the exception of operation 1, do not have a fixed point. This last one is an important property: as we shall see it confers a particular

character on the holonomy group. Rotation and symmetry operations, for example, do not belong to this group. But a symmetry operation with respect to a line accompanied by translation along the line may well belong to it.

7.6 Fundamental Polyhedron

We have now found that the properties of locally Euclidean space in a simply-connected domain $\mathscr{D}$ are the same as those for Euclidean space. This completes the third stage in our exploration of space. Now we ask a further fundamental question: what is the largest simply-connected domain $\mathscr{D}$ that we can find? This task will be the fourth stage in the exploration of locally Euclidean spaces, and in undertaking it we shall come to the limits of the latter.

To give the answer we introduce the concept of a fundamental polyhedron. The *fundamental polyhedron* $\mathscr{P}$ of a locally Euclidean space is composed of the ensemble of points $\underline{X}$ in the Euclidean space where the development is made which are nearer to $\underline{P}$ (arbitrarily chosen) than to its homologous points $\underline{Q}, \underline{R},\ldots$ (Fig. 7.4).

This polyhedron is thus bounded by plane faces $\mathscr{F}_Q, \mathscr{F}_R,\ldots$ in the case of three-dimensional space and by straight lines in the two-dimensional case.

The polyhedron is convex, for if $\underline{X}_1$ and $\underline{X}_2$ are two points in $\mathscr{P}$, they are on the same side of any face, and this is also true for all points $\underline{X}_3$ situated on the line $\underline{X}_1\underline{X}_2$.

It is possible to show that $\mathscr{P}$ has a finite number of faces. The argument essentially is based on the fact that the distance between a point and its homologous point is greater than or equal to a "critical" distance.

Lastly, the fundamental polyhedron's faces are homologous in twos: to every face $\mathscr{F}$ there corresponds (exactly one) face $\mathscr{F}'$ such that given a point $\underline{N}$ in $\mathscr{F}$ there is a unique point $\underline{N}'$ in $\mathscr{F}'$ such that $\underline{N}$ and $\underline{N}'$ are the developments of the same point $\underline{M}$ in the locally Euclidean space. Since the proof is short, we will give it: Consider the points $\underline{Q}, \underline{R},\ldots$ actually being used to construct the polyhedron relative to $\underline{P}$ and the corresponding displacement $D_Q, D_R,\ldots$ The inverse operation D_Q^{-1} transforms $\underline{Q}$ into $\underline{P}$ and $\underline{P}$ into $\underline{S} \neq \underline{Q}$, otherwise the middle point of $\underline{PQ}$ would be invariant. To these two points $\underline{Q}$ and $\underline{S}$ there correspond two faces $\mathscr{F}_Q$ and $\mathscr{F}_S$ of the polyhedron and the points on those two faces are homologous, in twos, since we go from a point in $\mathscr{F}_Q$ to the homologous point in $\mathscr{F}_S$ by the operation $D_S = D_Q^{-1}$.

The fact that the faces of the fundamental polyhedron are homologous in pairs shows that, if we construct the fundamental polyhedron relative to points $\underline{Q},\underline{R},\ldots$, (homologous to $\underline{P}$), we obtain a "pavement" - or "tiled pattern" for two dimensions - regular in Euclidean space, with paving stones - or tiles - with all of them close-packed.

The displacements that transfer from one polyhedron face to its homologous face are in fact the generating operations of the holonomy group. Every point $\underline{Q},\underline{R},\ldots$, developed from $\underline{M}$, can be reached in journeying from polyhedron to polyhedron: every crossing of a polyhedron can have a starting point and a finishing point, the latter being homologous to the former one through the generating operation which transports the fact of the starting point into the one of the finishing point.

The number of generating operations for the holonomy group is therefore finite, since the fundamental polyhedron has a finite number of faces.

In the cylinder example, the fundamental polygon is a strip of Euclidean plane, with width equal to the circumference of the cylinder, and of infinite length.

We now return to the main question to find the largest simply-connected domain $\mathscr{D}$ in locally Euclidean space around an arbitrary point M. One answer is given by that domain of which the development Δ in Euclidean space is the fundamental polyhedron $\mathscr{P}$ relative to point P, developed from M. In fact, if we take a Δ larger than $\mathscr{P}$, we encroach on one or several other polyhedra packing Euclidean space, and there would be points in locally Euclidean space with several developed points. $\mathscr{D}$ will therefore not be simply-connected.

Evidently this development Δ from $\mathscr{D}$ is not the only possibility. We could derive an infinite variety as follows: let a given domain δ overlap one or several polyhedra neighbouring $\mathscr{P}$, and remove from $\mathscr{P}$ the domain δ' homologous to δ; this operation can be repeated at will (Fig. 7.5), as long as we take care that the region so bordered will stay simply-connected.

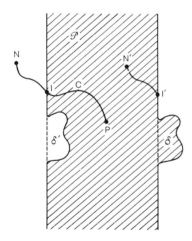

Fig. 7.5. The last but one stage in the exploration of locally Euclidean spaces

We will end this discussion with a remark on the developments. It can be shown that by developing the points of a locally Euclidean space we cover Euclidean space entirely once and once only. This requires certain restrictions: 1) the metric should be everywhere regular, i.e. with ds^2 positive definite; 2) the space should be a countable union of compacta. Such a space is then said to be *normal*. These conditions are quite acceptable, physically, for the space which we inhabit. They are satisfied, for example, by a cylindrical surface. There are other developable surfaces that do not have an everywhere regular metric: the apex of a cone, or the surface represented by the tangents to a twisted curve. If the apex of a cone is truncated, the metric is everywhere regular, but the surface is no longer compact.

The development of a cylinder onto a Euclidean plane covers the whole plane once, as we have seen. On the other hand, the development of a cone covers it entirely an infinite number of times; and, a further example: the development of the surface swept out by the tangents to a circular helix covers the plane twice, except for the inside of a circle larger than the base of the cylinder, where there is no covering.

If we assume that a space satisfies the above two restrictions, then we deduce that a packing of fundamental polyhedra entirely fills Euclidean space, *ad infinitum*.

7.7 Representing Locally Euclidean Space in Euclidean Space

Having reached the conclusion of the descriptions of developments, holonomy groups and fundamental polyhedra, we have at our disposal a useful tool for depicting locally Euclidean spaces with ease. Once we have packed Euclidean space using fundamental polyhedra whose characteristics correspond to the nature of the locally Euclidean space under consideration, we can plot (Fig. 7.5) a curve C that connects P̲ to some point N̲ in Euclidean space. This curve corresponds to a certain path Γ in locally Euclidean space and is the development of it. If C leaves the fundamental polyhedron relative to P̲ at a point I̲, the exterior portion I̲N̲ may be replaced by its homologue I̲'N̲' inside 𝒫, a section which is the development of the same section of Γ as I̲N̲. Through the generating operations of the holonomy group we may, therefore, reduce the development C of any exploratory path Γ in locally Euclidean space to a representation contained entirely inside the fundamental polyhedron relative to the origin of the path Γ.

Once more we can return to the cylinder model. A helix linking two points on the cylinder and making n turns will be represented by n + 1 elements of parallel straight lines zigzagging across the strip that represents the fundamental polygon.

The method therefore consists of "packing" Euclidean space using fundamental polyhedra and considering as identical those points in this space which, distributed among these various polyhedra, correspond one to another through the operations of the holonomy group. We can therefore, additionally, replace a point outside a selected polyhedron by its homologue inside, and in doing so concentrate the representation of locally Euclidean space to the interior of a single fundamental polyhedron.

7.8 The Various Types of Locally Euclidean Space

Thanks to this now unlimited means of exploration we can set out on the fifth and final stage in the study of locally Euclidean space: what structures will we encounter in journeying freely through those spaces?

Everything depends on working out the possible shapes of the fundamental polyhedra and, finally, the various possibilities for the generating operations of the holonomy group. This is a problem in group theory and the details will not concern us, just the main results. These latter depend critically on the dimensionality of space.

Two-dimensions - We have the identity operation; furthermore the only generating operations that satisfy the condition that they be discontinuous displacements without fixed points are: translations T and translations T_S accompanied by a symmetry operation with respect to a straight line parallel to the translation.

Using these operations we can construct five kinds of locally Euclidean spaces:

1) *Operation* 1: This is a trivial case: the fundamental polyhedron is the entire Euclidean plane. The surface of a corrugated sheet would give us, for example, a model embedded in three-dimensional Euclidean space. From all points of view, including the topological one, the locally Euclidean space in this case is identical to Euclidean space, and, in particular, it is simply-connected.

2) *Operations* 1 *and* T: We return once again to our cylindrical surface. The space is open, except in the direction orthogonal to the sides of the fundamental polygon where it is closed.

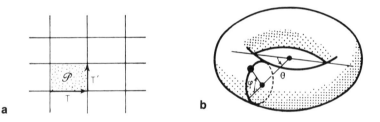

Fig. 7.6a, b. Locally Euclidean toroidal space

3) *Operations* 1, T *and* T': Two translations T and T' comprise the generating operations. The fundamental polygon $\mathscr{P}$ is a rectangle (Fig. 7.6a). This space is closed and has a surface area equal to $\mathscr{P}$. It is not possible to construct a model simply by embedding a surface in three-dimensional Euclidean space. But it is possible to represent it by a cylindrical torus whose metric is $ds^2 = d\theta^2 + d\varphi^2$, θ and φ being two angular coordinates for a point on the surface of the torus (Fig. 7.6b). We can easily imagine this space being homeomorphic with a rectangle: cut a tire along a small meridian circle, then along the larger equator and then stretch it out over a plane. One translation corresponds to the equatorial cut, the other to the meridional cut.

4) *Operations* 1 *and* T_S: Here we are going to define T_S thus: translation $\underline{XY}$ and a reflection with respect to the line D (Fig. 7.7a). The fundamental polygon $\mathscr{P}$ is again an infinitely long strip, as in the second case. Space is therefore open, except in one direction. But now an intriguing novelty is apparent; the space is not oriented. We see this by considering an asymmetric figure - refer to the little flags in Fig. 7.7a which we move along two different closed paths. One stays within $\mathscr{P}$, such as $\underline{ABA}$, or likewise $\underline{ACE}$ which arrives at $\underline{E}$, a point homologous to $\underline{A}$ and therefore identical to $\underline{A}$ and not affecting the flag. The other path $\underline{AFGE}$ leaves $\mathscr{P}$; the exterior section $\underline{FGE}$ can be replaced by $\underline{HIA}$ by the operation T_S^{-1}; this operation inverts the flag and brings it back to $\underline{A}$ in a position that cannot be matched with the departure position.

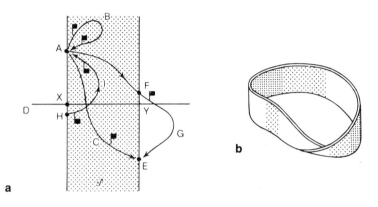

Fig. 7.7a, b. Moebius' locally Euclidean space

This locally Euclidean space does not have an embedded representation, but it is homeomorphic with the Moebius strip (Fig. 7.7b).

Straight lines in this space have very bizarre properties. Lines parallel to D are closed and of length 2XY except that contiguous with D which has a length of XY only. A straight line that is not parallel to D cuts itself infinite times: starting at $\underline{0}$ (Fig. 7.8a) we trace within $\mathscr{P}$ the half straight line $\underline{0}1.\underline{1}2.\underline{2}$... and its counterpart $\underline{0}a.\underline{a}b.\underline{b}c$... This straight line cuts itself between $\underline{0}$ and $\underline{1}$, $\underline{1}$ and $\underline{2}$,..., following the homeomorphic scheme sketched in Fig. 7.8b.

Direction is not absolute: in displacing us along the straight line $\underline{1}0ab$, we pass twice through the same point $\underline{p}$ in two different directions. Only the directions parallel or perpendicular to D are "invariable".

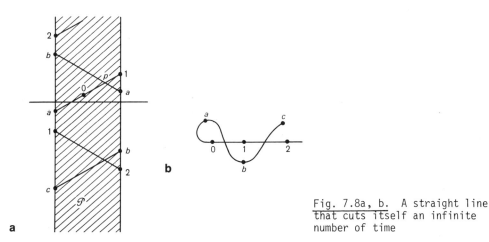

Fig. 7.8a, b. A straight line that cuts itself an infinite number of time

5) *Operations* 1, T, T_S: The fundamental polygon is a rectangle as in case 3, so space is closed and finite. But, as in case 4, it has no orientation (Fig. 7.9a). It is homeomorphic to a Klein bottle (Fig. 7.9b), which is obtained by connecting the top of a bottle via a hole in the body from the interior. Like the Moebius strip, the Klein bottle has only one side and, though closed, has neither an interior nor an exterior.

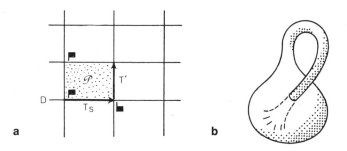

Fig. 7.9a, b. Klein's locally Euclidean space

It can be shown that no other combinations of generating operations without invariant points exist. Therefore, there are a total of five types of locally Euclidean space, homeomorphic with a plane, a cylinder, a torus, a Moebius strip, and a Klein bottle. Two are closed (torus, Klein) and two have no orientation (Moebius, Klein) and one is simply-connected (plane).

Three dimensions - In three-dimensional space the generating operations are 1, translations T, and helical displacements H. It can be shown that in three dimensions there are 18 different types of locally Euclidean spaces: 4 open orientable, 4 open non-orientable, 6 closed orientable and 4 closed non-orientable. One of them is simply-connected and identical to Euclidean space.

The fundamental polyhedra close-pack the Euclidean space in which the developments are made. The simplest cases are those in which these polyhedra are: slabs with parallel sides, rectangular prisms, rectangular parallelepipeds; these respectively correspond to one, two or three translations in the holonomy group.

This last space has a finite volume. Topologically it is equivalent to a hypertorus, a topological direct product of three circles. It is remarkable that closed spaces exist which should be endowed with the Euclidean metric and are orientable - in other words, actually close to the usual Euclidean space, but with finite volume; there are even six of these spaces.

An example of the polyhedron corresponding to helical displacements is the regular octahedron: passing from one face to the opposite face requires translation and rotation through $60°$. But this polyhedron cannot be a fundamental one because you cannot close-pack Euclidean space with regular octahedra. We shall see later that this is possible, on the other hand, with spherical space.

We end by stating a theorem: a normal space, in the above sense, locally Euclidean and simply-connected, is identical to Euclidean space.

Apparently the structure of locally Euclidean spaces may be very complex. This glimpse shows that, if observations demonstrate, via the model universes of General Relativity, that the space we live in has a zero Riemann-Christoffel tensor (i.e. zero curvature, as we will see later) then it needs not necessarily be Euclidean, and therefore infinite. A zero-valued Riemann-Christoffel tensor is, as we have already said, a pointwise property. Our space could very well at the same time have zero curvature and a finite volume. This contrasts with one current view that zero curvature implies infinite space. This would be true if space were simply-connected, but there is no proof that it is.

8. Locally Non-Euclidean Spaces

We are now going to consider a space in which the metric is defined by equation (7.1), but this time we shall let the Riemann-Christoffel tensor be arbitrary rather than being zero as in the preceding chapter.

Thus we tumble into the jungle of Riemannian spaces, spaces of extraordinary richness. The locally Euclidean spaces, which are a special case thereof, have already given us a glimpse of this complexity; and the complexity is enhanced, as we shall see, because the curvature is non-zero.

In fact, we shall limit this study to *regular Riemannian spaces*. These are spaces such that the neighbourhood of each point could be represented regularly and continuously by a certain neighbourhood of a point in Euclidean space with at least the same number of dimensions. Thus, Riemannian 2-space, defined by embedding an ellipsoid in Euclidean 3-space, is regular. The same is not so, however, for the embedding of a cone or the surface created by the tangents to a twisted curve, because at the top of the cone and on the twisted curve the conditions of regularity and continuity are violated.

These restrictions are justifiable because our aim is the study of the 3-space of our universe, space which, at every point, effectively permits the neighbourhood to satisfy the preceding conditions.

As in the previous chapter we proceed here also to explore space via successive steps, using exact and convenient representations in the customary Euclidean space, valid for successively greater domains.

8.1 First Order Representation

We will take a point $\underline{M}_o$ with coordinates u_o^i in a Riemannian space RS whose metric is given by the fundamental tensor g_{ij}, functions of the u^i:

$$ds^2 = g_{ij} du^i du^j. \tag{8.1}$$

In Euclidean space ES make a point $\underline{P}_o$ correspond to $\underline{M}_o$, with a reference frame of base vectors $\underline{e}_i$ such that

$$\underline{e}_i \cdot \underline{e}_j = (g_{ij})_o,$$

the values of g_{ij} at point $\underline{M}_o$.

To a point $\underline{M}$ infinitely close to $\underline{M}_0$ in RS we let point $\underline{P}$ correspond in ES such that

$$\underline{P}_0 \underline{P} = d\underline{P} = \underline{e}_i \, du^i. \tag{8.2}$$

Thus the neighbourhood of $\underline{M}_0$ is represented to first order in ES, which is called the *tangent Euclidean space* at $\underline{M}_0$; this is accomplished with the help of a *metric* known also as the *Euclidean tangent* metric at $\underline{M}_0$. This representation enables the concepts of vectors, tensors, and coordinate transformations to be extended to Riemannian space.

8.2 Second Order Representation

We consider again the first order representation, with the additional step of imposing at $\underline{P}_0$ a system of general coordinates such that the Christoffel symbols at $\underline{P}_0$ should be equal to those $(\Gamma_j{}^i{}_k)_0$ of the RS at $\underline{M}_0$. These symbols can be calculated from the g_{ij} and their first derivatives. For this it is sufficient to write

$$d\underline{P} = \underline{e}_i \left[du^i + \frac{1}{2}(\Gamma_j{}^i{}_k)_0 \, du^j \, du^k \right]. \tag{8.3}$$

In this way we represent the neighbourhood of M_0 to second order in ES, which we shall call the *osculating Euclidean space* at $\underline{M}_0$; this is done thanks to a *metric* that we shall call *osculatrix* at $\underline{M}_0$.

This representation is extremely important: now we can extend to Riemannian space all the operations and concepts of tensor calculus. We shall now give those that will be especially useful later on.

Absolute differential - Consider a vector with contravariant components v^i, functions of the coordinates. Its absolute differential is the vector $\nabla \underline{v}$ with contravariant components, denoted by ∇v^i, resulting from the own variations dv^i of the vector and from the changes of the vector due to the general coordinates used, $\omega_h{}^i v^h$; thus

$$\nabla v^i = dv^i + \Gamma_k{}^i{}_h \, v^h \, du^k \tag{8.4}$$

after replacing $\omega_h{}^i$ by its value (7.4).

Covariant derivative - The covariant derivative of a vector with components v^i is a tensor whose components $(\nabla v)_k^i$, written $\nabla_k v^i$, are given by

$$\nabla_k v^i = \partial_k v^i + \Gamma_k{}^i{}_h \, v^h. \tag{8.5}$$

Here, too, the first term on the right gives the proper change of the vector as a function of the coordinates u^i, whereas the second term gives the variations due to the coordinate system itself.

Divergence of a tensor - The divergence of a tensor T^{ij} is a vector whose components v^i, written $\nabla_j T^{ij}$, are given by

$$v^i = \nabla_j T^{ij} = \partial_j T^{ij} + \Gamma^i_{jm} T^{mj} + \Gamma^k_{km} T^{im}.$$

Parallelism - The concept of parallelism can also be introduced to Riemannian space through the absolute differential; two infinitely close vectors with components $v^i(u^1,...)$ and $v^i(u^1 + du^1,...)$ are parallel if the absolute differential is zero: $\nabla v^i = 0$.

Geodesic curve - Finally, the concept of a geodesic curve should be introduced; a geodesic curve is a curve along which the tangent vector remains parallel to itself as it is carried along the curve. Defining this curve by the parametric equations $u^i(t)$, the tangent vector has components $v^i = du^i/dt$ and the condition for parallel transport $\nabla v^i = 0$ may be written

$$\frac{d^2 u^i}{dt^2} + \Gamma^i_{kh} \frac{du^k}{dt} \frac{du^h}{dt} = 0. \tag{8.6}$$

These equations constitute a system of second-order differential equations that enable us to set up geodesic curves in a Riemannian space. We can now have access to the remarkable properties of these curves, which we shall do in the next step of exploration.

8.3 Development Along a Curve

We have stayed so far in the first two stages of exploration, infinitely close to $\underline{M}_0$ in RS. We are now going to venture freely, but with the restriction of confining the exploration within the infinitesimal neighbourhood of an arbitrary curve L starting from $\underline{M}_0$. This is obtained by making a second-order representation in ES, step-by-step, of the points $\underline{M}$ of L. We thus derive a curve C, said to be the development of L in ES. We will call this the *Euclidean contact space along* L. In the vicinity of C, the metric of RS is conserved in the sense that the metric of ES has the same coefficients g_{ij} and the same first-order derivatives as the coefficients of the metric of RS.

As a result of this conservation property and thanks to the representation used here, to two parallel vectors in ES at two points of C there are two corresponding parallel vectors of RS at two points of L, for their absolute differentials are zero. Hence a geodesic curve of RS develops along a straight line in ES, in accordance with the above mentioned definition of the geodesic.

There is a very important consequence: a geodesic curve achieves an extremum of length for a curve joining two points in RS, since its development achieves this in ES and since the metric is conserved.

Furthermore, for two "sufficiently close" points in RS there passes one and only one geodesic that has the minimum length, and if consideration is limited to a "sufficiently small" domain containing the two points, only one geodesic passes through them.

Finally, we can consider the change in a vector displaced along a curve L of RS in integrating the change of its representation along curve C, the development of L in ES. In particular, if we demand that this change in ES should be zero, we say the vector has undergone *parallel transport* along L in RS. We shall show later that the final position of a vector made to undergo parallel transport in RS depends on the path taken, in contrast to the result for ES.

In order to illustrate the definitions and properties encountered in this chapter we may still refer to a model of Riemannian 2-space obtained by embedding a surface in Euclidean 3-space. For locally Euclidean spaces we selected a circular cylinder which would be very suitable here, too, although for greater generality we take the surface of an ellipsoid (Fig. 8.1a). Through a point $\underline{M}_o$ of this RS two coordinate lines u_1 and u_2 pass. The tangent ES is given by considering a tangent plane to the ellipsoid at $\underline{M}_o$ and defining there a frame by the point $\underline{P}_o$ in contact with $\underline{M}_o$ and the two base vectors fused with the base vectors of the natural frame at $\underline{M}_o$. The osculating ES is obtained by taking for the coordinate lines at P_o the orthogonal projections onto the plane of the coordinate lines u_1 and u_2. These tangent and osculating ES give two representations, in ES, of the neigbourhood of $\underline{M}_o$, to first and second orders respectively.

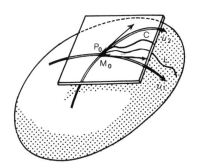

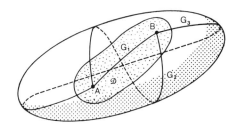

Fig. 8.1a, b. Models of Riemannian 2-space

To get a representation by development along a curve L of the ellipsoid it is sufficient to cause the ellipsoid to roll without slipping on the plane considered and along the curve L. The point of contact traces on the plane a curve C, developed from L. Now imagine that some pattern is carved into the ellipsoid so that it gets impressed on the plane in the vicinity of C and all along C as the ellipsoid is pressed against the plane. The resulting print is the second order representation on the Euclidean plane of a narrow strip of Riemannian space.

Geodesics are obtained by causing the ellipsoid to roll without slipping along straight lines on the plane. They are readily traced by stretching a frictionless wire on the ellipsoidal surface between two points $\underline{A}$ and $\underline{B}$ (Fig. 8.1b). The "direct" route G_1 gives the shortest geodesic. But there are others that are longer, such as G_2, that also give a local minimum length, and others still, like G_3, that give a maximum. If $\underline{A}$ and $\underline{B}$ are "sufficiently close", only one geodesic gives the absolute minimum length, but if $\underline{A}$ and $\underline{B}$ are, for example, at either end of a principal axis of the ellipsoid, there are two geodesics, and if they are at either poles of an ellipsoid of revolution, there are infinitely many geodesics: all the meridian lines of the ellipsoid. If attention is restricted to a small area $\mathscr{D}$, only one geodesic G_1 passes through $\underline{A}$ and $\underline{B}$. Our example shows that, as in the case of locally Euclidean spaces, problems of the simply-connectedness, the locality, are also posed by Riemannian spaces. In what follows we shall frequently need to assume that the elements considered are enclosed in a "sufficiently small" domain.

In order to illustrate conveniently the parallel transport of a vector we take a sphere instead of an ellipsoid. Parallel transport, from an equatorial point $\underline{A}$ to the north pole $\underline{N}$, of a vector $\underline{v}$ pointing north and taken along the shortest geodesic G_1, gives, with the method of development, a vector $\underline{v}_1$. But if one takes another route to reach $\underline{N}$, a different vector results: thus, if we paralleltransport $\underline{v}$ first along the equator to $\underline{B}$, at longitude λ, we get $\underline{v}'$; then, following geodesic G_2 to reach $\underline{N}$, we get a vector $\underline{v}_2$ making an angle λ with $\underline{v}_1$.

The Euclidean contact space along some curve provides a convenient representation to second order of Riemannian space in the vicinity of that curve. The metric is reproduced there with the same coefficients and first derivatives, but not, however, the second derivatives or those of higher order. We may not extend this representation along a curve to a zone wider than that neighbouring a curve, i e. to a representation along a surface of dimension greater than one. Hence it is impossible to impress a finite domain of the ellipsoid considered above on the plane and conserve the metric. If it were otherwise the metric on the ellipsoid would be Euclidean, like that of the plane, over a finite domain, which is not the case. This might happen in exceptional examples of Riemannian space: for instance, if the ellipsoid were replaced by a cylinder, and then we fall back again on locally Euclidean spaces.

But by introducing some simple concepts, in the sense of concepts that are physically clear, we arrive at an interesting representation that enables us to obtain a convenient and exact model of a Riemannian space provided that it belongs to a particular and remarkable class of non-Euclidean spaces, those with constant curvature that are just used in cosmology. This is the geodesic representation.

8.4 Geodesic Surfaces

From now on we limit ourselves to three-dimensional space, which does not matter too much as that is what we are interested in.

A *geodesic surface* GS at a point $\underline{M}$ in Riemannian 3-space is the surface comprising all the geodesics passing through $\underline{M}$ and tangent to a plane element at $\underline{M}$.

A *totally geodesic surface*, TGS, is a surface that is geodesic at all its points. (This concept was introduced by HADAMARD). A TGS is equivalent to a plane in Euclidean space. So, if a geodesic curve is tangent to a plane element of the TGS at one of its points, it is entirely located in the surface as a result of the definition.

Furthermore, every geodesic curve which has two sufficiently close points in a TGS is entirely located in it. Indeed, let $\underline{A}$, $\underline{A}'$ be these two points on the geodesic curve G. From $\underline{A}$, geodesic curves start that entirely fill the TGS, at least locally in a finite domain. Therefore, one of these, G', passes through $\underline{A}'$. But, through two points belonging to a sufficiently small region, there passes only one geodesic G. G is therefore identical to G' and hence is entirely located on the TGS.

This property is equivalent to that whereby a straight line through two points on a plane is entirely located in the plane.

The existence of the TGSs in Riemannian spaces is exceptional. We say that, if a Riemannian space is such that at a point $\underline{M}$ and tangent to any plane element at that point, there passes a TGS, then that Riemannian space satisfies the *axiom of the plane* at $\underline{M}$.

Schur's first theorem - If the axiom of the plane is verified for two sufficiently close points, we can find a representation in Euclidean space such that any geodesic curve can be represented by a straight line, and vice versa.

The proof may be demonstrated by projective geometry and although it is rather long we give it here because the theorem is an important one and also because the proof familiarises us with the structure of Riemannian spaces that satisfy the axiom of the plane.

> We consider a point $\underline{M}$ in RS and the two geodesic curves MA and MB passing through the points A and B where the axiom of the plane is satisfied (Fig. 8.2). Now we suppose that $\underline{M}$, $\underline{A}$ and $\underline{B}$ are contained in a sufficiently small region.
>
> The geodesic curve passing through $\underline{A}$ is defined by the components x,y,z of its unit tangent vector at A. The other at B is defined by x',y',z'. In total, six numbers are tied to $\underline{M}$, which reduces in fact to four ratios. But $\underline{M}$ depends on only three variables. So there exists a relation among x,y,...,z', which we shall now derive.
>
> A TGS passing through $\underline{A}$ is governed by a linear equation, homogeneous in x,y and z which may be written

$$f_1 + mf_2 = 0,$$

where the functions f_1, f_2 are linear and homogeneous forms in x,y,z and where m is some parameter. Similarly for a TGS through $\underline{B}$ we may write

$$f_1' + m'f_2' = 0.$$

Now we shall consider a TGS containing the geodesic curve $\underline{AB}$. Associated with it there are two linear and homogeneous (in x,y,...,z') equations, and therefore there is a relation between m and m'. To specify this we consider four TGSs containing $\underline{AB}$ and also consider the plane element at $\underline{A}$ perpendicular to $\underline{AB}$, where these four TGSs cut in four directions. Carry out the same operation $\overline{at}$ $\underline{B}$.

One can then show that the anharmonic ratio of the four directions at $\underline{A}$ is the same as that at B. Therefore a homographic relation exists between m and m' and, by a suitable choice of three of the four TGSs, this relation can be written m = m', i.e. $f_1/f_2 = f_1'/f_2'$.

This relation between x,y,...,z' may also be written

$$f_1 = \rho f_1' \equiv X$$
$$f_2 = \rho f_2' \equiv Y$$

where ρ is a constant and where X and Y are given by definition. Now consider a linear homogeneous relation f_3 in x,y,z, independent of f_1 and f_2, and similarly f_3' for x',y',z', independent of f_1' and f_2'. Then set

$$f_3 = Z$$
$$\rho f_3' = T.$$

So, at each point $\underline{M}$ there correspond four quantities X,Y,Z,T whose ratios are obtainable from linear homogeneous expressions. Conversely, from the three independent ratios X/Y/Z/T we obtain x,y,...,z', and therefore a point M. Summarising these considerations, we have set up a coordinate system X,Y,Z,T to indicate the points $\underline{M}$ in a region of Riemannian space, sufficiently small, satisfying the axiom of the plane at $\underline{A}$ and $\underline{B}$.

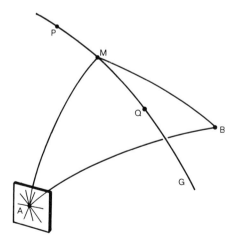

Fig. 8.2. Schur's first theorem

Let us now consider some geodesic curve G passing through some point M̲ and the two TGSs passing through G and A̲ and through G and B̲. To these correspond two linear equations, homogeneous in X,Y,Z and in X,Y,T, respectively. Thus G is defined by two equations of the first degree in X,Y,Z,T.

Conversely, two linear equations in X,Y,Z,T may be written in the form of an equation in X,Y,Z and an equation in X,Y,T, to which there correspond two TGSs, one through A̲ and the other through B̲. These two TGSs intersect each other, following a curve which is a geodesic by virtue of the axiom of the plane at A̲ and B̲; in effect, let us consider two points P̲ and Q̲ on that curve of intersection and the geodesic curve P̲Q̲. This last curve is entirely in each of the two TGSs, therefore the intersection is a geodesic curve.

If now we consider X,Y,Z,T as the homogeneous coordinates of a point m̲ in Euclidean 3-space, to the points M̲ of a geodesic curve G in RS there correspond the points m̲ of a straight line D̲ in ES - the equations being linear - and the converse is also true.

We have thus found a *geodesic representation* of RS for which any geodesic curve can be made to correspond to a straight line in ES, and vice versa.

At this stage this geodesic representation is rather formal and theoretical for our purposes, but we shall use it later in this form to describe the Riemann-Christoffel tensor more precisely, and subsequently we shall cast it into a convenient and readily-used form.

Schur's second theorem - If the axiom of the plane is satisfied at two sufficiently close points then it is satisfied everywhere. This is a rather remarkable theorem; it gives a first glimpse of the homogeneity of the structure of Riemannian spaces that satisfy the axiom of the plane. The proof is easy if the geodesic representation is used.

Consider any point M̲ in RS and any GS Γ passing through M̲. From the geodesic representation, to Γ there corresponds a plane Π passing through P̲, the image of M̲. Consider any point Q̲ of Π and the straight lines on Π passing through Q̲. Corresponding to these, by the geodesic representation, there is a point N̲ of Γ and a GS entirely contained in Γ. Therefore Γ is a GS at N̲, and therefore a TGS because N̲ is arbitrary - we may, in effect, reach every point N̲ of Γ sufficiently close to M̲ by a geodesic curve that emanates from M̲. Finally, the axiom of the plane is satisfied at M̲, and therefore everywhere.

It may seem somewhat strange that a property which is valid for two points only leads to validity everywhere. This arises because the satisfaction of the axiom of the plane at one point implies in fact special characteristics, at the very least over the whole spatial domain touched by the TGSs passing through the point considered.

8.5 The Riemann-Christoffel Tensor

The geodesic representation allowed us, in the fourth stage of exploration, to represent the whole of a domain of Riemannian space; this domain is limited nevertheless by the restriction of being "sufficiently small". Now we shall use this representation to find the form of the Riemann-Christoffel tensor in spaces that satisfy the axiom of the plane.

We take ds^2 of RS (8.1), a function of coordinates u^i (i = 1,2,3). By the geodesic representation, to u^i there correspond coordinates x, y, z in ES, and so a ds^2 that is a function of x, y, z. From the requirement that curves in RS that have extrema in their arc-lengths should be straight lines in ES, we get relations between the g_{ij} and their first derivatives $\partial_k g_{ij}$; these relations can be expressed as functions of the variables x, y, z. Considering again the variables u^i we get (we shall not dwell on the details of the calculation) the relations

$$R_i{}^h{}_{lk} = K(\delta^h_l g_{ik} - \delta^h_k g_{il}) \tag{8.7}$$

where the Riemann-Christoffel tensor $R_i{}^h{}_{lk}$ is expressed by equation (7.10) as a function of the Christoffel symbols and their derivatives, where K is a constant and δ^j_i is the Kroneker symbol.

To bring out the nature of the relations (8.7) we shall use, instead of an arbitrary coordinate system u^i, a simpler system of orthogonal coordinates at the point under consideration; in other words, such that $g_{ij} = \delta^j_i$. Then we find

$$\left.\begin{array}{l} R_i{}^h{}_{lk} = 0 \\ R_i{}^h{}_{hi} = -R_i{}^h{}_{ih} = K \end{array}\right\} \tag{8.8}$$

(without summation over h here).

Thus it appears that the components of the Riemann-Christoffel tensor are equal to a constant K, or they are zero. Relation (8.8) is crucial because it brings us to the concept of space with constant curvature, the first examples of which, historically, were Riemann's spherical space and Lobatchevski's hyperbolic space.

Spaces whose Riemann-Christoffel tensor satisfies relations (8.7) or (8.8) are named, by generalising (7.10), *locally non-Euclidean spaces*, the term *locally* encompassing the restrictions that we have so far introduced into our arguments. If K > 0, the space is *locally spherical* or of *constant positive curvature*; if K < 0, it is *locally hyperbolic* or of *constant negative curvature*; and finally, if K = 0, we recover the locally Euclidean space of *zero curvature*.

8.6 Riemannian Curvature

We now adopt a locally non-Euclidean space (LNES) and consider the parallel transport of a vector v around an infinitesimally-small closed path with area $d\mathscr{S}$(Fig. 8.3a). The element $d\mathscr{S}$ can be calculated through the intermediary of the tangent Euclidean space. In constructing the development of this loop in ES we obtain an arc PP' and a vector v' that finishes up in position w' parallel to v' (Fig. 8.3b). But in LNES the final position w does not correspond precisely to the initial position v. There has been a rotation dφ, in the plane of the loop, given by

dφ = Kd𝒮. (8.9)

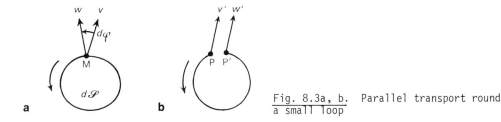

Fig. 8.3a, b. Parallel transport round a small loop

We shall not demonstrate that this formula holds for the general case, but only in the particular case of a model derived by embedding a sphere of radius R in Euclidean 3-space (Fig. 8.4a). The loop Γ represented by a circle with the infinitesimally-small spherical polar angle α develops as the arc of a circle C with radius Rα (Fig. 8.4b). A parallel-transported vector will have turned an angle dφ = πα², and since the area of the cycle is d𝒮 = π(Rα)², we find

dφ = d𝒮/R².

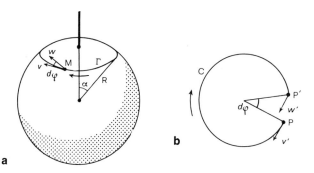

Fig. 8.4a, b. Parallel transport for the case of a spherical model

On the other hand, one can show by taking a simple coordinate system in the space being considered and by making the calculations (see next chapter) that the Riemann-Christoffel tensor has components - in addition to the null ones - with the value K = 1/R². Therefore dφ = Kd𝒮.

The rotation is of order $d\mathscr{S}/R^2$, whereas the break PP' of the developed loop is $(d\mathscr{S})^{3/2}/R^2$, of a higher order; for practical purposes we can consider loop C to be closed.

For a locally spherical space LSS the rotation has the same sense as the direction along the loop, whereas the opposite is true in locally hyperbolic space LHS. And, sure enough, it is zero for a locally Euclidean space LES.

By analogy with the ordinary theory of surfaces the constant K, which is inversely proportional to the square of the radius of the model sphere, is called the *Riemannian curvature*. The Riemann-Christoffel tensor may also be called the *curvature tensor*.

It can be shown that the rotation (8.9) is independent of the orientation in LNES of the loop. This is expressed by saying that LNES is *isotropic*. A third theorem due to Schur establishes that a Riemannian space which is isotropic at all points has constant curvature. The directional effect encompassed in the first part of the theorem somehow propagates by degrees in order to create a constant curvature at every point.

We end this section on Riemannian curvature with an interesting example: consider a *geodesic triangle* ABC formed by three geodesic lines AB, BC, CA and with interior angles α, β, γ. From the axiom of the plane it is possible to see that, as a result of the geodesic representation, these three lines are in the same TGS. Starting from A, paralleltransport a vector v_1, tangent at A to AB, to position v_1' at point B. At B, let us turn it, in the TGS, by an angle $\pi - \beta$ into v_2 and then paralleltransport that to C where we turn it by $\pi - \gamma$, and lastly we go to A and turn by $\pi - \alpha$. The net result is that the vector has been rotated round the cycle ABCA by an angle

$$K\mathscr{S} + 3\pi - \alpha - \beta - \gamma$$

where $\mathscr{S}$ is the area of the cycle given by integrating its elements $d\mathscr{S}$. Since two neighbouring loops $d\mathscr{S}_1$ and $d\mathscr{S}_2$ may be combined into a single larger loop, relation (8.9) allows extension to the finite cycle ABCA. We thus obtain *Gauss' theorem*, according to which the sum of the angles in a triangle differs from π:

$$\alpha + \beta + \gamma = \pi + K\mathscr{S}. \tag{8.10}$$

This result furnishes the most direct and physical method of determining "in the laboratory" the curvature of space, simply by surveying techniques. If the universe were sufficiently curved, the difference between $\alpha + \beta + \gamma$ and π would be measureable and would give K, the curvature, and its sign. Unfortunately the radius of curvature of the universe is thought to be at least of order 10^{28} cm. If a triangle having the dimensions of the orbit of the Earth round the Sun were to be used, a precision of 10^{-25} arc sec must be achieved in measuring angles. If the orbit of the Sun round the galactic centre were to be used, a precision of 10^{-6} arc sec would do, and radio astronomers can already achieve 10^{-3} arc sec. But, to make this measurement would take more than 100 millions of years...

Another measure of curvature often used is the *scalar curvature*:

$$S = R_i^{\ i} \tag{8.11}$$

where $R_i^{\ j}$ is the *Ricci tensor*

$$R_{ij} = R_i^{\ k}{}_{kj} \tag{8.12}$$

obtained by contracting the Riemann-Christoffel tensor.

At this point we recall that contracting the indices of a tensor, $t_i^{\ j}{}_{kl}$ for example, consists of summing the terms where a superior index j is equal to an inferior index i, to give the contracted tensor

$$c_{kl} = t_i^{\ i}{}_{kl}.$$

It can be shown that

$$S = n(n-1)K \tag{8.13}$$

where n is the number of dimensions.

8.7 General Properties of Locally Non-Euclidean Spaces

In this chapter we finally arrived at a class of Riemannian 3-spaces, the locally non-Euclidean spaces endowed with several remarkable properties:

1) They satisfy the axiom of the plane;
2) They permit a geodesic representation;
3) Their Riemann-Christoffel tensor has essentially one constant component K;
4) Their Riemannian curvature is constant;
5) They are isotropic, as defined in Section 6 of this chapter.

It may be shown that these spaces enjoy a sixth important property:

6) They satisfy the axiom of free mobility in the sense that they allow the same group of isometric transformations as Euclidean space. Another way of saying this is that arbitrary shapes can be displaced without deformation by movements that have the same generality as the motions of a solid body in Euclidean space - translation, rotation, symmetry operations, helical displacements. This property is of physical importance. If it were not so, a solid object could find that, as a result of the geometry of space, it could not move. For example, it is impossible on the surface of an ellipsoid with three different axes to move a part of that surface without deformation. However, this operation is possible on a sphere.

Equally, it can be shown that all six of the above properties involve each other.

It is this set of properties that gives such a particular place to locally non-Euclidean spaces, as opposed to Riemannian spaces generally, and confers on them such an important cosmological role.

8.8 The Various Types of Locally Non-Euclidean Spaces

In the previous chapter we saw that there are four different types of locally Euclidean 2-dimension spaces and eighteen of 3-dimension spaces. For locally non-Euclidean spaces the situation is frightfully complicated and has not actually been entirely elucidated.

The various types are defined here, too, by the various generating operations that constitute the holonomy group. One can associate fundamental polyhedra to these operations. However, the polyhedra are located in spaces of constant curvature and may only be visualised in a schematic fashion.

For the case of locally spherical 2-spaces we may, however, return to our standard model of the surface of a sphere embedded in Euclidean 3-space. In this case the fundamental polygons should cover the sphere, in the same way that polygons cover the Euclidean plane for locally Euclidean 2-spaces. In fact, in a general way, it can be shown that the fundamental polyhedra of a locally non-Euclidean space of Riemannian curvature K pack the simply-connected non-Euclidean space with the same curvature K and the same dimensionality. The next chapter is concerned with these simply-connected spaces. Here we shall only say that the whole surface of a sphere embedded into Euclidean space is a model of simply-connected space with constant positive curvature.

To elaborate a little on the various types of locally non-Euclidean spaces, we now consider separately the spherical ($K > 0$) case and the hyperbolic ($K < 0$) case.

Locally spherical spaces - If the number of dimensions n is even, only two types exist: *spherical* space and *elliptical* space. For example, for $n = 2$, the first type is given by the surface of a sphere embedded in Euclidean space and the second by the same model when, in addition, diametrically opposed points are considered to be identical.

In the spherical case, the fundamental polygon is the entire sphere. Space is finite, of area $4\pi/K^2$, orientable and simply-connected.

In the elliptical case the fundamental polygon is a hemisphere limited by two sides, the two semi-circles $\underline{PEP'}$ and $\underline{P'E'P}$ (Fig. 8.5). Space is still finite and with area $2\pi/K^2$, half that of spherical space. It is non-orientable and not simply-connected. Effectively, if a "flag" is carried from $\underline{A}$ to $\underline{A}$ by a path that does not leave the fundamental polygon $\mathscr{P}$ related to $\underline{A}$, it returns with the same orientation. But if we leave $\mathscr{P}$ to arrive at $\underline{A'} \equiv \underline{A}$ along $\underline{AEA'} \equiv \underline{AE} + \underline{E'A}$, the flag will return with the opposite orientation. Equally we see that the closed circuit $\underline{AEA'}$ cannot be continuously reduced to a circle of zero radius; the space is not simply-connected. This geometry is the same as the well known one of the *elliptical plane*.

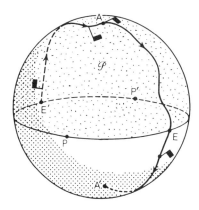

Fig. 8.5. Model of elliptical space of two dimensions

If the number of dimensions n is uneven, there exists, in addition to the spherical and elliptical types, an infinite sequence of varieties, all closed. For n = 3 an example is furnished by the case in which the holonomy group includes as the generating operation a helical displacement consisting of a translation and a rotation through 60°. The fundamental polyhedron is a regular octahedron: every face corresponds to an opposing face. We already noted that these octahedra cannot pack Euclidean space. But in case of space of positive curvature, by applying Gauss' theorem, the sum of the angles of each triangular face is greater than π, and, for certain values of the sides of the octahedron with respect to $1/\sqrt{K}$, the angles at the apex of the octahedra may be large enough to fill spherical 3-space.

Another known example is that where the holonomy group includes a helical displacement with a rotation of 36° and where the fundamental polyhedron is a regular dodecahedron bringing the face ABCDE on the opposite one A'B'C'D'E' (Fig. 8.6). This case involves 120 generating operations and lets us see the extreme complexity of the displacements that could be generated by these operations.

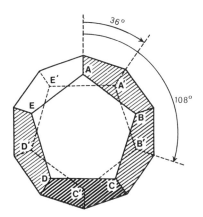

Fig. 8.6. Locally spherical or hyperbolic dodecahedric 3-spaces

Locally hyperbolic spaces - Here the situation is still more complex. In two dimensions open and closed forms exist. Only two theorems are known:

1) All closed surfaces of Euclidean space, other than the sphere, the elliptical plane (which are homeomorphic to locally spherical spaces), the torus and the Klein bottle (which are homeomorphic to locally Euclidean spaces), are homeomorphic to locally hyperbolic spaces. So, a sphere with handles is homeomorphic to a space with constant negative curvature.

2) Any open surface of Euclidean space is homeomorphic to a locally hyperbolic space, for example a hyperboloid, with or without handles.

In the three-dimensional case, no one has yet successfully determined the shapes. Only some examples are known, of which some are also closed. Thus we cite the case where the fundamental polyhedron is a dodecahedron (as in the case of positive curvature), but, however, the holonomy group has a helical displacement with a rotation of $108°$ instead of $36°$ (Fig. 8.6).

Here also, in contrast to the usual widely-held opinion, a three-dimensional space of negative, even constant, curvature is not necessarily infinite. An infinity of them with finite volume exists. On the other hand, all spaces of constant positive curvature are closed.

9. Spherical and Hyperbolic Spaces

Among the jungle of Riemannian spaces, we have selected out the remarkable class with constant curvature: locally non-Euclidean spaces which still are of extraordinary variety, as we have seen at the end of the previous chapter. Among locally Euclidean spaces, showing also a large variety of forms, the one which is simply-connected, the Euclidean space, appears as the simplest example. Similarly here, among locally non-Euclidean spaces those which are simply-connected will also have the barest essentials, and these are the easiest to study. This chapter is devoted to two of the simply-connected locally non-Euclidean spaces, simply called *non-Euclidean spaces*. One, with positive curvature K, is the *spherical space*, and the other, with negative curvature K, is the *hyperbolic space*. We shall not speak again of space with zero curvature, Euclidean space.

These three spaces are undoubtedly the simplest Riemannian spaces, the most widely used in cosmology. In this chapter we will limit our attention, as a rule, to the three-dimensional case.

9.1 Geodesic Representation

The most important task that we ought to undertake is to find a coordinate system such that the ds^2 (8.1) represents a space with constant curvature; and, at the same time, to understand through the intermediary of a representation which provides an exact model, the physical significance of these coordinates; this is in order to visualise the spaces that we shall consider.

Quite naturally, we start with the geodesic representation of the previous chapter, by which every geodesic of non-Euclidean space has a corresponding straight line in the Euclidean space where we make the representation. In the representation the metric cannot be that of the Euclidean space, otherwise one is representing a space that is also Euclidean. The metric is obtained by using the *Cayley distance* as the distance between two points. Let $\underline{P}$, $\underline{P}'$ be any two points on any straight line D in the geodesic representation of two points $\underline{M}$, $\underline{M}'$ of a geodesic line G of the non-Euclidean space. We consider some fixed sphere Σ with radius iR, called the *absolute*. Let $\underline{N}$, $\underline{N}'$ be the points where D cuts Σ. The distance s between the points $\underline{M}$, $\underline{M}'$ is given by *Cayley's formula*

$$s = \frac{R}{2i} \text{Log}(PP'NN') \tag{9.1}$$

where $(PP'NN')$ is the anharmonic ratio of the four points $\underline{P},\underline{P'},\underline{N},\underline{N'}$ of the Euclidean space.

We must now show that with these definitions we effectively obtain a ds^2 leading to constant curvature. For simplicity, we are content here with the two-dimensional case and suppose R to be purely imaginary. The sphere is therefore a circle Σ (Fig. 9.1). But the demonstration can be made more generally. We introduce polar coordinates θ, ρ for $\underline{P}$ and consider $\underline{P'}$ infinitesimally near to $\underline{P}$. Then

$$ds \approx \frac{R}{2i} \text{Log}\left(1 + \frac{PP'}{N'P} - \frac{PP'}{NP}\right) \approx \frac{R}{2i}(d\rho^2 + \rho^2 d\theta^2)^{1/2} \frac{NN'}{PN.PN'}.$$

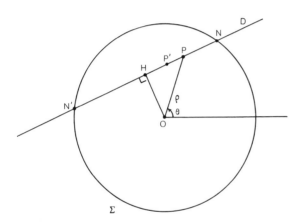

Fig. 9.1. Geodesic representation of hyperbolic 2-space

The denominator of the last factor is the power of $\underline{P}$ with respect to Σ, equal to $-R^2 - \rho^2$, whereas

$$NN' \equiv 2NH$$

is expressed from $\underline{NO}$ and $\underline{OH}$. Ultimately, we get

$$ds^2 = \frac{d\rho^2 + \rho^2(1 + \rho^2/R^2)d\theta^2}{(1 + \rho^2/R^2)^2}. \tag{9.2}$$

The fundamental tensor has these components:

$$\left.\begin{array}{l} g_{11} = (1 + \rho^2/R^2)^{-2} \\ g_{22} = \rho^2/(1 + \rho^2/R^2) \\ g_{12} = g_{21} = 0 \end{array}\right\}. \tag{9.3}$$

Using these components it is possible to calculate in turn the Christoffel symbols, the Riemann-Christoffel tensor, the Ricci tensor, the scalar curvature and, ultimately, the Riemannian curvature K. The calculations are given in the appendix to this chapter. They lead finally to the relation

$$K = 1/R^2. \tag{9.4}$$

Therefore the Riemannian curvature is constant, the space represented is locally non-Euclidean and, in particular, here it is locally hyperbolic, since R is pure imaginary and therefore K < 0.

Furthermore, it is simply-connected. In effect, s is real for points P, P' interior to the absolute Σ. On the other hand, as P' comes towards the absolute from the interior, the distance from P, interior to the absolute, to P', tends to infinity. These two properties clearly entail that the domain represented by the interior of the absolute be simply-connected. Therefore the space is non-Euclidean and, here in particular, hyperbolic.

Our demonstration, limited to two dimensions, can be extended to the three-dimensional case. The geodesic representation of hyperbolic space fills the inside of a sphere Σ of real radius $iR = \sqrt{-K}$. Here the geodesic lines are represented by sections of straight lines inside Σ and the TGSs by portions of planes also inside Σ. The ds^2 is given by

$$ds^2 = \frac{d\rho^2 + \rho^2(1 + K\rho^2) da^2}{(1 + K\rho^2)^2} \tag{9.5a}$$

where

$$da^2 = d\varphi^2 + \sin^2\varphi \, d\theta^2 \tag{9.5b}$$

is the angle element corresponding to ordinary spherical coordinates ρ, θ, φ in the Euclidean space where the representation is made. θ and φ are the normal angular variables of a coordinate system with origin at O, the image of a certain point C of non-Euclidean space, which are used to locate an arbitrary point P, the image of M, while ρ is a radial variable related to the distance separating M from C along the geodesic line CM. For ρ small, that is, in the vicinity of C, θ and φ are the same in the Euclidean space used to represent the non-Euclidean space, as in the non-Euclidean space itself because the element ds^2 (9.5) reduces to the ds^2 of

Euclidean space, $ds^2 = d\rho^2 + \rho^2 da^2$.

Therefore the geodesic representation at point $\underline{C}$ is effectively a Euclidean tangent space.

For the geodesic line $\underline{CM}$, θ and φ are fixed and the distance $s = \underline{CM}$ is expressed as a function of ρ by integration of (9.5):

$$s = \int_0^r d\rho/(1 + K\rho^2). \tag{9.6}$$

For $\rho = \sqrt{-K}$, the integral diverges; the points represented by the absolute are truly infinitely far from $\underline{C}$.

Properties of hyperbolic space - This geodesic representation of hyperbolic space shows us that hyperbolic space is open; points $\underline{N}$ and $\underline{N}'$ are at infinity on the line D; any geodesic line G extends to infinity in both directions.

Through two points $\underline{M}$, $\underline{M}'$ there passes one geodesic line and one only. This geodesic is the minimum path between $\underline{M}$ and $\underline{M}'$.

Two geodesic lines contained in the same TGS may or may not intersect. So, if we are given in a TGS a geodesic G and a point $\underline{P}$ outside G, then within the TGS we may find three types of geodesics G' through $\underline{P}$: those that cut G at a finite distance, those that cut G at infinity and those that do not cut G at all. The geodesic representation of these features is easy to construct. In this way we derive the well-known result of LOBATCHEVSKI's hyperbolic geometry according to which it is possible to draw, through a point, two "straight lines" parallel to a "straight line".

A triangle has its sides parallel in pairs when the vertices are on the absolute.

For spherical space, the curvature K is positive, $R = \sqrt{K}$ is real and the radius of the absolute iR is purely imaginary. The geodesic representation is of little interest in this case, although the ds^2 given by (9.5) is still valid.

9.2 Central Representation

The most convenient representation for spherical space is derived by changing the radial variable used in the geodesic representation (9.5):

$$\omega = \tan^{-1}(\rho\sqrt{K}). \tag{9.7}$$

The ds^2 then becomes

$$ds^2 = (d\omega^2 + \sin^2\omega \, da^2)/K. \tag{9.8}$$

With this expression for ds^2 the distance of a point from the origin is simply

$$s = \int_0^\omega d\omega/\sqrt{K} = \omega/\sqrt{K}. \tag{9.9}$$

For K > 0 we immediately obtain a model in the two-dimensional case; we say model and not representation here because we are not going to employ Euclidean 2-space to construct it. The ds^2 (9.8) is quite simply the same as on the surface of a sphere Σ' of radius R' = $1/\sqrt{K}$ in Euclidean space, where ω is the angle at the centre serving to reference the image point P with respect to a point O on the sphere, while θ is the position angle of P with respect to O (Fig. 9.2). In this model the geodesic lines are great circles on the sphere Σ'. It is obvious that the space is simply-connected; it is therefore spherical 2-space.

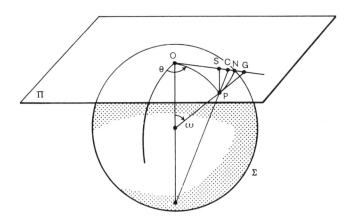

Fig. 9.2. Model of spherical 2-space

In considering a series of spheres with radii $\sqrt{(R'^2 - t^2)}$, where t represents the fourth variable in Euclidean 4-space, varying in the interval (-R',R'), we obtain a model of spherical space in three dimensions, also simply-connected.

The quantity $1/\sqrt{K}$ is called the *radius of curvature* of non-Euclidean space, because of the model that we have described. The radius of curvature is *real* in spherical space, *pure imaginary* for hyperbolic space.

This representation of the ds^2 (9.8) will be called the *central representation*. This representation is a Euclidean tangent space at point C of the non-Euclidean space, the image of C being the origin O in the Euclidean space.

This representation therefore conserves the angles at O and, for symmetry reasons, at every point P of the model given by the spherical surface of Fig. 9.2. In tracing great circles on this sphere Σ', we recover Gauss' theorem concerning the sum of the angles in a triangle.

Properties of spherical space - Some further properties of spherical space can be deduced immediately from this model

- a TGS passing through M is represented by a sphere such as Σ' in Fig. 9.2;
- the geodesics are closed and of length $2\pi/\sqrt{K}$;
- any TGS is closed and of area $4\pi/K$ - later we will work out the volume of the spherical space;

- two geodesic lines lying within the same TGS necessarily intersect at two distinct points;
- the geodesic lines leaving a given point all intersect at a same point, termed the *antipole* of the point considered, after a distance of $\pi/\sqrt{K}$;
- from a point $\underline{M}$ it is possible to draw only one geodesic line that intersects another geodesic G perpendicularly, unless $\underline{M}$ is at a distance $\pi/2\sqrt{K}$ from geodesic G, and in this case there are an infinite number of perpendicular geodesic lines. In this case we speak of $\underline{M}$ as a *pole* of geodesic line G; it is located in a TGS passing through G. G is the *equator* of the point $\underline{M}$ located in a TGS passing through $\underline{M}$. Point $\underline{M}$ thus has an infinite number of equators which form a surface known as the *equatorial surface of* $\underline{M}$, whose points are at a distance $\pi/2\sqrt{K}$ from $\underline{M}$.

As a result of the model of spherical 2-space given by Fig. 9.2, we may readily obtain a geodesic representation. For this it suffices to project the points $\underline{P}$ of the sphere Σ' in $\underline{G}$ from its centre on the plane Π tangential at $\underline{O}$ to Σ'. Thus any geodesic of the space is represented by a straight line. Furthermore, we recover the ds^2 (9.5), since $OG = K^{-1/2} \tan \omega$. But the same point $\underline{G}$ of Π corresponds to two diametrically opposed points of Σ': in fact, this geodesic representation is the one of elliptical two-dimensional space.

9.3 Other Representations

With the help of other changes of variable carried out on the radial variable we can obtain other representations, each of which presents a particularly interesting feature, suitable for some problem or other. Changes of the angular variables are of no great interest because the directional properties of non-Euclidean space at a point $\underline{M}$ are the same as those at point $\underline{P}$ in Euclidean space; the major property is that of isotropy, in the sense that all directions are equivalent. This is not the same as for the radial variable since, as we have shown in equation (9.6), this is sensitive to effects associated with the curvature of space.

Conformal representation - The *conformal representation* conserves the value of angles in the Euclidean space that is used to represent non-Euclidean space. It is obtained, in the simple case illustrated in Fig. 9.2, by making a stereographic projection in $\underline{C}$ onto a plane Π of the points $\underline{P}$ of sphere Σ'. The geodesic lines, which are great circles on Σ', are projected as circles.

The change of variable giving the conformal representation is

$$u = 2K^{-1/2} \tan(\omega/2) \tag{9.10}$$

and the corresponding value of ds^2 is

$$ds^2 = (du^2 + u^2 da^2)/(1 + Ku^2/4)^2. \tag{9.11}$$

A particularly interesting construction leads to the conformal representation of hyperbolic 2-space from its geodesic representation. Taking Σ as the absolute circle of this geodesic representation, we consider some sphere Σ" centred on the axis $\underline{IH}$ of Σ and the apex $\underline{H}$ of a cone with base Σ; this cone encircles Σ" along the circle Γ" (see Fig. 9.3). A straight line D of the geodesic representation projects with respect to $\underline{H}$ into a circle Γ on Σ". We now make a stereographic projection of sphere Σ" with respect to $\underline{J}$, one of the intersections of $\underline{IH}$ with Σ". The circles Γ and Γ" thus project into circles C and Δ.

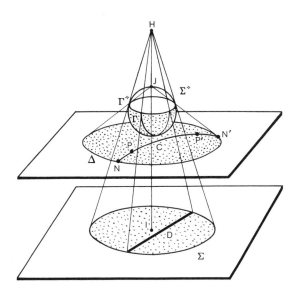

Fig. 9.3. Constructing the conformal representation of hyperbolic 2-space

The result of these operations is to transform the absolute Σ and a line D into a circle Δ and another circle C intersecting Δ at a right angle; the interior of Σ corresponds to the interior of Δ. A geodesic of hyperbolic 2-space is therefore represented by a circle cutting the absolute at a right angle at $\underline{N}$ and $\underline{N}'$. By repeating in detail the transformations it is possible to show that the changes of variable (9.7) and (9.10) have effectively been realised, and therefore the representation is conformal.

It can be shown that the distance between two points, in hyperbolic space, that have representations $\underline{P}$ and $\underline{P}'$ on the circle C is

$$s = (-K)^{-1/2} \log (PP'NN') \qquad (9.12)$$

where (PP'NN') is the anharmonic ratio of points $\underline{P}, \underline{P}', \underline{N}, \underline{N}'$.

The sum of the angles of a triangle is less than π; it may even be zero if the three vertices are on the absolute.

By making an inversion with respect to a point of Δ *Poincaré's conformal representation* is obtained. In this the absolute is a straight line and the geodesics are semicircles centred on this line.

Normal representation - If, as radial variable, we use the distance of a point $\underline{M}$ from an origin $\underline{C}$ in non-Euclidean space, we obtain a *normal representation*. Let us reconsider the model shown in Fig. 9.2: the representation is obtained by developing the arc $\underline{OP}$ of sphere Σ' according to the segment $\underline{ON}$ of the plane Π. The radial variable is

$$v = \omega/\sqrt{K} \tag{9.13}$$

and the ds^2 in the normal representation is

$$ds^2 = dv^2 + \frac{\sin^2(v\sqrt{K})}{K} da^2. \tag{9.14}$$

Sinusoidal representation - If the radial variable is taken as

$$w = \sin\omega/\sqrt{K} \tag{9.15}$$

then ds^2 assumes the very simple form

$$ds^2 = \frac{dw^2}{1 - Kw^2} + w^2 da^2. \tag{9.16}$$

In the particular case corresponding to Fig. 9.2 w is simply the length OS where $\underline{S}$ is the projection of point $\underline{P}$ on the plane Π. Because a sine term enters equation (9.15) we shall call this the *sinusoidal representation*.

If K is negative, w will require us to use the hyperbolic sine and $\sqrt{|K|}$.

Reduced sinusoidal representation - The most useful representation that we shall use in the model universes treated in the third part of this book - because it gives the simplest formulae - is obtained by selecting a reduced radial variable

$$r = |\sqrt{K}| w. \tag{9.17}$$

Then ds^2 is written

$$ds^2 = R^2 \left(\frac{dr^2}{1 - kr^2} + r^2 da^2 \right) \tag{9.18}$$

with

$$\left. \begin{array}{l} k = 1 \text{ if } K > 0 \\ k = -1 \text{ if } K < 0 \end{array} \right\}$$

and $\quad R = 1/\sqrt{|K|}$. $\tag{9.19}$

This representation, which we shall term the *reduced sinusoidal* representation, uses a reduced radial variable without dimension r. In place of the Riemannian curvature K, we use the constant k = ± 1 to represent the sign of K and a radius of curvature R, always taken positive, which is obtained from the modulus of K.

With the ds^2 (9.18) it is easy to calculate the volume of a sphere centred on $\underline{C}$ as a function of r:

$$V = \int_0^r \int_0^{2\pi} \int_{-\pi/2}^{\pi/2} \sqrt{g} \, dr \, d\theta \, d\varphi \qquad (9.20)$$

where g is the determinant derived from the components of the fundamental tensor. Here

$$g = \begin{vmatrix} \dfrac{R^2}{1-kr^2} & 0 & 0 \\ 0 & R^2 r^2 & 0 \\ 0 & 0 & R^2 r^2 \sin^2 \varphi \end{vmatrix}. \qquad (9.21)$$

In hyperbolic space we find

$$V = 2\pi R^3 [r(1+r^2)^{1/2} - \text{Log}\{r + (1+r^2)^{1/2}\}] \qquad (9.22a)$$

and in spherical space

$$V = 2\pi R^3 [\sin^{-1} r - r(1-r^2)^{1/2}]. \qquad (9.22b)$$

The radius of the sphere is given by

$$s = \int_0^r R \, dr/(1-kr^2)^{1/2} . \qquad (9.23)$$

For hyperbolic space we obtain

$$s = R \sinh^{-1} r \qquad (9.24a)$$

and for spherical space

$$s = R \sin^{-1} r. \qquad (9.24b)$$

For s small with respect to R, we find $V \simeq 4\pi s^3$, as in Euclidean space. But if s is not small, V is, in the hyperbolic case, greater than the value given by the Euclidean formula, whereas in the spherical case V is smaller.

As the distance s is a quantity of physical importance, we give in Fig. 9.4 its behaviour in units of R, s/R, as a function of the radial variable in the reduced sinusoidal representation. For small values of r we have $s \simeq Rr$. For hyperbolic space, s increases less rapidly than r, but increases without limit as r varies from 0 to ∞.

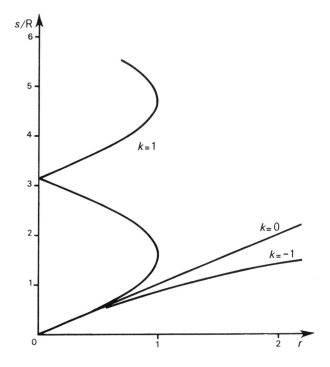

Fig. 9.4. Distance as a function of the radial variable in the reduced sinusoidal representation

For spherical space, s increases more rapidly than r and for r = 1 we reach the equatorial surface of $\underline{C}$ at radius s = πR/2. Further excursion results in r decreasing from 1 to 0 and the antipole of $\underline{C}$, at s = πR, is reached at r = 0.

If we pass through the antipole, we approach the departure point from the opposite direction. Following the normal convention, to ensure that r remains positive we can change θ and φ to $\theta + \pi$ and $-\varphi$. Then, in a new cycle of variation with r going from 0 to 1 and then back to 0, we return from the antipole to $\underline{C}$, having covered a total distance of s = 2πR and having recrossed the equatorial surface at s = 3πR/2 - whence the two sinusoidal arches in Fig. 9.4. In the spherical case, r varies only through the interval (0,1) and s/R is defined to within an integral number of times 2π; this results from the fact that, if we go from $\underline{C}$ along a geodesic line, we return after having made an excursion that is an integral number of times the distance 2πR, as shown by the model of Fig. 9.2.

We note here that to one given value of r there correspond two values of s/R in the interval (0,π), and these are symmetrical with respect to the value $\pi/2$, corresponding to the equatorial surface. In this sense the sinusoidal representation is improper for the spherical case. This constitutes a slight inconvenience that we must not lose sight of, but it is not of any importance if we study only the portion of space interior to the equatorial surface.

Using the formulae derived above we can calculate the volume V_s of spherical space. We obtain it when the radius of the sphere is πR, r having varied from 0 to 1 and back to 0; (9.22b) thus leads to

$$V_s = 2\pi^2 R^3. \tag{9.25}$$

The volume of elliptical space would be $V_E = \pi^2 R^3$.
The surface S of the sphere centred on $\underline{C}$ and with radius defined by r is simply

$$S = \int_0^{2\pi} \int_{-\pi/2}^{\pi/2} R^2 r^2 \sin\varphi\, d\theta\, d\varphi = 4\pi R^2 r^2 \tag{9.26}$$

whereas the circumference of a circle centred on $\underline{C}$ and of the same radius is

$$C = \int_0^{2\pi} Rr\, d\theta = 2\pi Rr. \tag{9.27}$$

Expressing r as a function of the radius s by (9.24) we get, for small values of s, $S \simeq 4\pi s^2$ and $C \simeq 2\pi s$, as in the Euclidean case. For large values of s the surface and circumference are, for hyperbolic space, larger than the Euclidean values, whereas for spherical space they are smaller.

9.4 Appendix

In this appendix we give the main line of argument for calculating the Riemannian curvature of a 2-space whose fundamental tensor and ds^2 are given as a function of the two coordinates. The main aim of the appendix is not so much to show that the curvature is given by (9.4) as to serve as an example of tensor calculus, so useful in cosmological problems, and to show the complexity that, in reality, is hidden beneath the apparently simple and compact tensor notation; the complexity is such that it is sensible even if we stick to two dimensions only.

We have to show that the metric defined by equations (9.2) and (9.3) leads to spaces of constant curvature given by (9.4). For reasons of simplicity we prefer to use, which is allowed, the central representation of the metric given by the ds^2 (9.8), where we do not know yet that we must replace K^{-1} by R^2.

Then the fundamental tensor is given by

$$g_{11} = R^2,\ g_{22} = R^2 \sin^2\omega,\ g_{12} = g_{21} = 0, \tag{9.28}$$

ω being the first coordinate and θ the second coordinate.

We devide the calculation into successive stages.

1. *Calculation of the determinant g of the fundamental tensor* - We simply have

$$g = \begin{vmatrix} R^2 & 0 \\ 0 & R^2 \sin^2 \omega \end{vmatrix} = R^4 \sin^2 \omega. \qquad (9.29)$$

2. *Calculation of the contravariant components g^{ij} of the fundamental tensor* - They are given by

$$g^{ij} = \alpha^{ij}/g \qquad (9.30)$$

where α^{ij} is the minor of the determinant g relative to the term ij. Whence

$$g^{11} = 1/R^2, \quad g^{22} = 1/R^2 \sin^2 \omega, \quad g^{12} = g^{21} = 0. \qquad (9.31)$$

3. *Calculation of the Christoffel symbols of the first kind Γ_{ijk}* - These are given by equation (7.8). As they are symmetric with respect to their outer indices and since the derivatives of the g_{ij} with respect to θ are zero, it suffices to calculate certain terms only:

$$\Gamma_{111} = 0, \quad \Gamma_{112} = 0, \quad \Gamma_{121} = 0, \quad \Gamma_{122} = R^2 \sin \omega \cos \omega, \quad \Gamma_{212} = -R^2 \sin \omega \cos \omega;$$

finally there remains

$$\Gamma_{122} = \Gamma_{221} = -\Gamma_{212} = R^2 \sin \omega \cos \omega \qquad (9.32)$$

all other terms Γ_{ijk} being zero.

4. *Calculation of the Christoffel symbols of the second kind $\Gamma_i{}^j{}_k$* - We obtain these by raising one index

$$\Gamma_i{}^j{}_k = g^{jh} \Gamma_{ihk}, \qquad (9.33)$$

an operation similar to that of lowering indices given by (7.5). The $\Gamma_i{}^j{}_k$ are zero, except possibly those where $\Gamma_{ijk} \neq 0$. Furthermore, they are also symmetrical with respect to their lower indices, whence

$$g^{12} \Gamma_{122} + g^{11} \Gamma_{112} = \Gamma_1{}^1{}_2 = 0 = \Gamma_2{}^1{}_1$$
$$g^{22} \Gamma_{122} + g^{21} \Gamma_{112} = \Gamma_1{}^2{}_2 = 1/\tan \omega = \Gamma_2{}^2{}_1$$
$$g^{11} \Gamma_{212} + g^{12} \Gamma_{222} = \Gamma_2{}^1{}_2 = -\sin \omega \cos \omega$$
$$g^{21} \Gamma_{212} + g^{22} \Gamma_{222} = \Gamma_2{}^2{}_2 = 0.$$

Therefore all the $\Gamma_i{}^j{}_k$ are zero except

$$\Gamma_1{}^2{}_2 = \Gamma_2{}^2{}_1 = 1/\tan \omega, \quad \Gamma_2{}^1{}_2 = -\sin \omega \cos \omega. \qquad (9.34)$$

5. *Calculation of the components $R_i{}^h{}_{lk}$ of the Riemann-Christoffel tensor* - These components are given by the first member of equation (7.10); being antisymmetric with respect to the last two indices, we have

$$R_i{}^h{}_{11} = R_i{}^h{}_{22} = 0.$$

It remains to calculate

$$R_1{}^1{}_{12} = -R_1{}^1{}_{21} = \partial_\omega \Gamma^1{}_{1\,2} - \ldots = 0$$
$$R_2{}^2{}_{21} = -R_2{}^2{}_{12} = \ldots = 0$$
$$R_1{}^2{}_{12} = -R_1{}^2{}_{21} = \ldots = \partial_\omega(1/\tan\omega) + 1/\tan^2\omega = -1$$
$$R_2{}^1{}_{12} = -R_2{}^1{}_{21} = \ldots = \partial_\omega(-\sin\omega\cos\omega) - (1/\tan\omega)(-\sin\omega\cos\omega) = \sin^2\omega.$$

Finally,

$$R_1{}^2{}_{12} = -R_1{}^2{}_{21} = -1, \quad R_2{}^1{}_{12} = -R_2{}^1{}_{21} = \sin^2\omega. \tag{9.35}$$

6. *Calculation of the covariant components R_{ij} of the Ricci tensor* - We obtain them by contracting the indices of the Riemann-Christoffel tensor in accordance with equation (8.12)

$$R_{11} = R_1{}^1{}_{11} + R_1{}^2{}_{21} = 1$$
$$R_{22} = R_2{}^1{}_{12} + R_2{}^2{}_{22} = \sin^2\omega$$
$$R_{12} = R_1{}^1{}_{12} + R_1{}^2{}_{22} = 0,$$

and since the Ricci tensor is symmetric, we also have $R_{21} = 0$.

7. *Calculation of the scalar curvature S* - Following (8.11) we contract the indices of the Ricci tensor, for which it is first necessary to raise an index. Then

$$S = g^{ij} R_{ji} = g^{11} R_{11} + g^{22} R_{22} = 2/R^2.$$

8. *Calculation of the Riemannian curvature K* - (8.13) then finally yields, for n = 2 dimensions

$$K = S/2 = 1/R^2. \tag{9.36}$$

Q.E.D.

Part III
Model Universes

We have summarised the elements of tensor calculus that we shall be using and we have also set out the major properties of spaces with constant curvature in the second part which was dedicated to geometrical aspects of the cosmological problem. We can now tackle the mechanistic aspects which, through the medium of General Relativity, will enable us to model the behaviour of the universe at large.

In this part, for simplicity, we replace the terms locally non-Euclidean, locally spherical, etc., by non-Euclidean, spherical, etc.

10. Uniform Relativistic Model Universes

10.1 The Equations of General Relativity

We shall start by setting down briefly the principles of Einstein's General Theory of Relativity. To describe phenomena occurring in the universe this theory uses a Riemannian 4-dimensional space, such that at each point ds^2 assumes a form, by a judicious choice of coordinates, in which the squared terms only appear with the signs -, -, -, +:

$$ds^2 = -(dx^1)^2 - (dx^2)^2 - (dx^3)^2 + (dx^4)^2. \tag{10.1}$$

This can be expressed by saying that ds^2 has *signature* - - - +. The value of ds^2 is therefore not necessarily positive. We mention in passing that Riemannian spaces with non positive ds^2 have structure very much more complicated than spaces with positive ds^2, such as those considered in the second part, and furthermore we are now working in four-dimensional space rather than three.

The asymmetry in the signature of ds^2 accounts for a fundamental physical phenomenon: among the four coordinates x^i necessary to describe the universe, three are the spatial coordinates x, y, and z, and one is the time coordinate t. Thus the name *spacetime* is given to this Riemannian space.

Though Special Relativity taught us that the spatial coordinates or the time coordinate may, in the Lorentz transformation, each be transformed into a linear combination of the spatial and time coordinates (and thereby blur the distinction between the notions of space and time), it is still a fact that this blending has intrinsic limitations. These arise because the ds^2 of spacetime has the signature - - - +. It is thanks to these limitations that the notions of space and time keep a certain reality all the same.

Furthermore, this ds^2 is such that, at every point, the ds^2 of the tangent Euclidean space is the same as that of Special Relativity, that is Minkowski spacetime

$$ds^2 = -dx^2 - dy^2 - dz^2 + c^2 dt^2 \tag{10.2}$$

where c is the speed of light, x, y, z are orthogonal spatial coordinates and t a time coordinate. This condition conveys a fundamental role to the speed of light and shows in particular that geodesic lines of zero length linking different coordinate points exist.

The foundation of General Relativity lies in the fact that the geometrical structure of spacetime is determined by its contents, both matter and energy:

(geometry) ⇔ (matter-energy). (10.3)

The geometry is defined by the components g_{ij} of the fundamental tensor, and the matter-energy content is described by the *energy-momentum tensor* T_{ij}. (10.3) is symbolically written

$$(g_{ij}) \Leftrightarrow (T_{ij}). \qquad (10.4)$$

The energy-momentum tensor is given by the theory of Spècial Relativity. Let us imagine that the universe is filled with a fluid of particles. To the movement of a particle in space as a function of time there corresponds a *worldline*, in other words a curve of spacetime defined by the ds^2 (10.2); this curve may be represented parametrically by four coordinates x, y, z, ct

$$x^i = x^i(s) \qquad (10.5)$$

as a function of the arc s of the curve.

The particle's speed may be described by the *unitary velocity (four-velocity)*, defined by the unit vector tangent to the worldline. This vector has components

$$u^i = dx^i/ds. \qquad (10.6)$$

If the fluid density is ρ, which may be a function of the coordinates x^i, the energy-momentum tensor of the matter is

$$T^{ij} = \rho u^i u^j. \qquad (10.7)$$

We note that the density ρ at a point is the *proper density*, that is the density measured in a spatial coordinate system such that, at the point under consideration, the matter is observed to be at rest.

If, furthermore, there is a pressure p in the fluid of particles, the energy-momentum tensor is

$$T^{ij} = (\rho + p/c^2)u^i u^j - (p/c^2)g^{ij}. \qquad (10.8)$$

This is the case of a *perfect fluid*, in other words a medium everywhere characterised by a scalar pressure p, a density ρ and a velocity $\underline{u}$. Examples are: a cloud of dust, a molecular gas, or a photon gas.

Now we shall consider an example. Suppose that in the neighbourhood of a point $\underline{0}$ of spacetime we have established a system $\mathscr{S}$ of coordinates x, y, z, ct such that the ds^2 of the tangent Euclidean space is given by (10.2), the values x = 0, y = 0, z = 0, t = 0 corresponding to point $\underline{0}$. A particle passing through point x = 0, y = 0, z = 0 at the instant t = 0, with zero velocity with respect to $\mathscr{S}$, has a worldline, passing through $\underline{0}$ and tangential to the axis of t at $\underline{0}$. More generally, if it passes through x = 0, y = 0, z = 0 at the instant t = 0 with velocity v along the x-direction, the worldline is tangential at $\underline{0}$ along a direction with coordinates vτ, 0, 0, cτ, where τ is timelike. The components u^i of the four-velocity are obtained by requiring that the vector (vτ, 0, 0, cτ)/l be of modulus 1 in the metric (10.2); the fundamental tensor of this metric is

$$g^{ij} = \begin{Vmatrix} -1 & 0 & 0 & 0 \\ 0 & -1 & 0 & 0 \\ 0 & 0 & -1 & 0 \\ 0 & 0 & 0 & 1 \end{Vmatrix} ; \tag{10.9}$$

l is a length obtained thus

$$l^2 = -v^2\tau^2 + c^2\tau^2.$$

The quantity τ/l is thus defined and yields

$$u^1 = \beta/\sqrt{1 - \beta^2}, \quad u^2 = 0, \quad u^3 = 0, \quad u^4 = 1/\sqrt{1 - \beta^2} \tag{10.10}$$

with β = v/c.

Now suppose that we have an ensemble of these particles such that at $\underline{0}$ they have a proper density ρ and a mean velocity v. Suppose, furthermore, that the random velocities create a pressure p at $\underline{0}$. Then the energy momentum tensor is

$$T^{ij} = \begin{Vmatrix} \left(\rho + \frac{p}{c^2}\right)\frac{\beta^2}{1-\beta^2} + \frac{p}{c^2} & 0 & 0 & \left(\rho + \frac{p}{c^2}\right)\frac{\beta}{1-\beta^2} \\ 0 & \frac{p}{c^2} & 0 & 0 \\ 0 & 0 & \frac{p}{c^2} & 0 \\ \left(\rho + \frac{p}{c^2}\right)\frac{\beta}{1-\beta^2} & 0 & 0 & \left(\rho + \frac{p}{c^2}\right)\frac{1}{1-\beta^2} - \frac{p}{c^2} \end{Vmatrix} . \tag{10.11}$$

Here we remark that as v → c, some terms of the tensor tend to infinity. In fact, no material particle can pass through the velocity c; the worldlines of particles passing through $\underline{0}$ are all inside a cone in hyperspace - a hypercone - defined by

$$ds^2 = 0, \tag{10.12}$$

usually called the *light cone*. If the z-dimension is suppressed, for example, this hypercone would be a cone of revolution about the t-axis (Fig. 10.1). The light cone divides the infinitesimal neighbourhood of spacetime around $\underline{0}$ into two regions: the interior contains lines through $\underline{0}$ called *timelike lines*, such as the worldlines, and the exterior contains *spacelike lines*. Lines through $\underline{0}$ located on the hypercone have

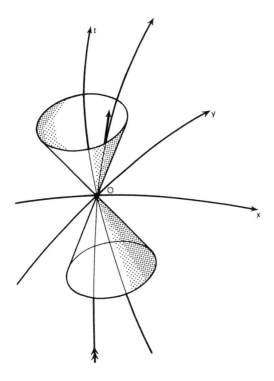

Fig. 10.1. Light cone and worldline

ds^2 zero; they correspond to points that move at the velocity of light, like photons.

From here it is sufficient for us to consider the case of a perfect fluid. The energy-momentum tensor is symmetric and the number of independent components reduces from 16 to 10. Furthermore, it has been defined in such a way that its divergence is zero:

$$\nabla_i T^{ij} = 0. \tag{10.13}$$

These four equations reduce the number of independent components to six.

Having described the energy-momentum tensor, we must now find, for writing out equation (10.4), a quantity that describes the geometrical properties of spacetime and satisfies the following conditions:

1) it is a function of the g^{ij};
2) it is a tensor;
3) it contains second order partial derivatives, like for every problem in mechanics;
4) it is symmetrical with respect to its indices;
5) its divergence is zero.

It can be shown that the most general tensor which satisfies these conditions is the tensor

$$S^{ij} = R^{ij} - \frac{1}{2}g^{ij}(S - 2\Lambda) \tag{10.14}$$

where R^{ij} is the Ricci tensor, S the scalar curvature and Λ an arbitrary *cosmological constant*. Then equation (10.4) leads to *Einstein's equations*:

$$S^{ij} = -\varkappa T^{ij}, \tag{10.15}$$

a system of ten equations with second order partial derivatives of the ten unknown functions g^{ij}, where $\varkappa$ is a constant. Einstein's equations are analogous to Poisson's equation linking the potential V and matter density ρ with the aid of the gravitational constant G

$$V = -4\pi G\rho,$$

and for this reason we call the g^{ij} the *gravitational potentials*. It is this analogy that guided Einstein and led him to give the value

$$\varkappa = 8\pi G/c^2$$

to constant $\varkappa$.

There exist among these ten equations four relations (since the divergence of S^{ij} is zero). Therefore, in fact, six equations remain for determining the ten g^{ij}. The four relations for g^{ij} that are indeterminate correspond to the degrees of freedom in the choice of the coordinate system or, said otherwise, to the various coordinate changes which can be made inside the spacetime, the intrinsic structure of which is effectively defined by six equations.

Having thus determined the geometrical structure of spacetime from its content it is necessary to determine the evolution of this content in order to solve the problems of mechanics completely. The solution is given by the *principle of geodesics*, according to which a material particle has for its worldline a geodesic in spacetime, which line is defined by the initial position and velocity of the particle. From this we can deduce its motion and hence solve all dynamical problems. Objects moving at the velocity of light - photons, ultrarelativistic electrons, high energy neutrinos - travel along geodesics of zero length. We note that later work has shown that this principle is a consequence of Einstein's equations.

Finally, we observe that as $G \to 0$, Special Relativity is recovered and, similarly, as $c \to \infty$, Newtonian Mechanics.

10.2 Dingle's Equations

Einstein's equations are, in their general form, practically insoluble. However, one can simplify the problem and use, in the spacetime, a system x^i of orthogonal coordinates. The ds^2 is then written thus:

$$ds^2 = D(dx^4)^2 - A(dx^1)^2 - B(dx^2)^2 - C(dx^3)^2 \qquad (10.16)$$

where A, B, C, D are functions of the four x^i. Here x^4 is the timelike coordinate. We shall follow closely McVITTIE's presentation, in which ds, following a usage that is becoming current, has the dimension of time.

We introduce the notation

$$A_i = \partial A/\partial x^i, \quad A_{ij} = \partial^2 A/\partial x^i \, \partial x^j, \ldots \qquad (10.17)$$

We can then calculate the 64 Christoffel symbols, 24 of which are zero in orthogonal coordinate systems; among the 40 others we get, for example,

$$\Gamma^1_{1\,1} = A_1/2A, \quad \Gamma^1_{2\,1} = A_2/2A$$
$$\Gamma^2_{1\,1} = -A_2/2B, \quad \Gamma^1_{2\,2} = -B_1/2A. \qquad (10.18)$$

Einstein's equations have simplified a little and lead to *Dingle's equations* of which three out of ten examples are given here:

$$-\varkappa c^2 D T^{44} = \frac{1}{2}\left(\frac{A_{22} + B_{11}}{AB} + \ldots\right) - \frac{1}{4}\left(\frac{A_2 B_2 + B_1^2}{AB^2} + \ldots\right) + \Lambda$$

$$\varkappa c^2 ABT^{12} = -\frac{1}{2}\frac{DC_{12} + CD_{12}}{CD} + \frac{1}{4}\left(\frac{C_1 C_2}{C^2} + \ldots\right) \qquad (10.19)$$

$$-\varkappa c^2 ADT^{14} = -\frac{1}{2}\frac{CB_{14} + BC_{14}}{BC} + \frac{1}{4}\left(\frac{B_1 B_4}{B^2} + \ldots\right).$$

These equations represent an indispensable step for the solution of cosmological problems.

10.3 Cosmological Solution

To pursue the solution of Einstein's equations further, in the cosmological case we may very happily introduce further justifiable simplifications. The first consists in recognising that from our observation point - the Earth - space appears to be isotropic. This *isotropy hypothesis* is reasonable enough: observations give no indication of anisotropy - the counts of galaxies and above all of radio sources give near enough the same results for all parts of the celestial sphere; the expansion follows the same pattern in various directions; and the 3 K background radiation is essentially isotropic. As a result of this, we can write ds^2 with a spherically symmetric spatial form and with an orthogonal time:

$$ds^2 = h(r,t)dt^2 - f(r,t)(dr^2 + r^2 da^2)/c^2 \qquad (10.20)$$

where r is a radial coordinate measured from the Earth and da^2 is the angular element (9.5b) corresponding to the angular coordinates θ and φ, and where h and f are not functions of θ and φ. Considering the particular case $h \equiv 1$, ds^2 can be written

$$ds^2 = dt^2 - e^{f(r,t)}(dr^2 + r^2 da^2)/c^2 \qquad (10.21)$$

where f is another function. Assigning the numerals 1, 2, 3, 4 to r, θ, φ, t, Dingle's functions are obtained:

$$A = e^f/c^2, \quad B = e^f r^2/c^2, \quad C = \frac{e^f r^2}{c^2}\sin^2\varphi, \quad D = 1. \qquad (10.22)$$

The second hypothesis is to suppose that the universe is filled with a perfect fluid. This *perfect fluid hypothesis* is equally justified: as has been clearly demonstrated in the first part of this book, one may consider galaxies as the molecules of a gas that fills space; and at the epoch when galaxies would not have existed, when the universe would probably have been very condensed, it would have been filled, as we shall see, by a photon gas which behaved like a perfect fluid. One may therefore use the energy-momentum tensor given by (10.8) where functions ρ and p describing the density and pressure appear.

With these hypotheses Dingle's equations where differentials ∂_2 and ∂_3 appear throughout, are zero. Then, after (10.19)

$$T^{12} = 0. \qquad (10.23)$$

The same holds true for T^{13}, T^{23}, T^{24} and T^{34} and there remain only five Dingle equations.

$T^{12} = 0$ leads to, after (10.8), $u^1 u^2 = 0$ because $g^{12} = 0$ and, with the help of similar equations, we see that

$$u^2 = u^3 = 0. \qquad (10.24)$$

Now, $u_i u^i = 1$, so

$$-e^f(u^1)^2/c^2 + (u^4)^2 = 1 \qquad (10.25)$$

and we are left with only four equations representing the Einstein equations; they correspond to indices 14, 44, 11 and a combination of 22 and 33.

$$\varkappa\left(\rho + \frac{p}{c^2}\right)u^4 u^1 = e^{-f}f_{rt} \qquad (10.26)$$

$$\varkappa\left[\left(\rho + \frac{p}{c^2}\right)(u^4)^2 + \frac{p}{c^2}\right] = -e^{-f}\left(f_{rr} + \frac{1}{4}f_r^2 + \frac{2}{r}f_r\right) + \frac{3}{4}\frac{f_t^2}{c^2} - \frac{\Lambda}{c^2} \qquad (10.27)$$

$$\varkappa\left[\left(\rho + \frac{p}{c^2}\right)(u^1)^2 + e^{-f}p\right]e^f = c^2 e^{-f}\left(\frac{1}{r}f_r + \frac{1}{4}f_r^2\right) - \left(f_{tt} + \frac{3}{4}f_t^2\right) + \Lambda \qquad (10.28)$$

$$\kappa p = c^2 e^{-f}\left(\frac{1}{2}f_{rr} + \frac{1}{2r}f_r\right) - \left(f_{tt} + \frac{3}{4}f_t^2\right) + \Lambda \qquad (10.29)$$

where, for example, f_{rt} signifies $\partial^2 f/\partial r \partial t$.

Thus, the hypothesis of isotropy with respect to the Earth and the perfect fluid hypothesis, in conjunction with the use of orthogonal coordinates, have reduced Einstein's ten equation to a mere five (10.25 to 10.29) which feature five unknown functions of r and t: u^1, u^4, f, ρ, p.

We are now going to look for a solution in the simplest possible way. Following McVITTIE, from (10.26) we may take $u^1 = 0$ if $f_{rt} = 0$, whence we deduce u^4 via (10.25). If we integrate $f_{rt} = 0$ with respect to r, then

$$f_t = 2\dot{R}/R$$

where R is an arbitrary function of t and $\dot{R}$ its derivative. This is another way of saying that $2\dot{R}/R$ is a constant with respect to r.

If we now integrate with respect to t, we get

$$f = 2\text{Log } R(t) + \chi(r) \qquad (10.30)$$

where $\chi(r)$ is an arbitrary function of r.

If, finally, we eliminate p between equations (10.28) and (10.29), ρ disappears because $u^1 = 0$, and Λ disappears, too. All calculations completed, there remains

$$f_{rr} - \frac{1}{r}f_r - \frac{1}{2}f_r^2 = 0.$$

Replacing f by (10.30) we find a differential equation in χ

$$\chi'' = \frac{1}{r}\chi' + \frac{1}{2}\chi'^2$$

which has the solution

$$\chi(r) = -2\text{Log}(1 + Cr^2/4) \qquad (10.31)$$

where C is a constant of integration.

Finally, with expressions (10.30) and (10.31), ds^2 (10.21) is written thus:

$$ds^2 = dt^2 - \frac{R(t)^2}{c^2} \cdot \frac{dr^2 + r^2 d\alpha^2}{(1 + Cr^2/4)^2}, \qquad (10.32)$$

and there remain only two Einstein equations, (10.27) and (10.29).

Then, lastly, making the change of radial variable $Cr^2 \to kr^2$ where the constant k= 1 if C > 0, -1 if C < 0 and 0 if C = 0, we obtain for ds^2 and the two remaining Einstein equations these final expressions:

$$ds^2 = c^2 dt^2 - R(t)^2 \frac{dr^2 + r^2 da^2}{(1 + kr^2/4)^2} \tag{10.33}$$

$$\frac{8\pi G}{c^2} p = - \frac{2\ddot{R}}{R} - \frac{\dot{R}^2}{R^2} - \frac{kc^2}{R^2} + \Lambda \tag{10.34}$$

$$8\pi G\rho = \frac{3}{R^2}(\dot{R}^2 + kc^2) - \Lambda \tag{10.35}$$

where we have again given ds the dimension of length. From equations (10.33) to (10.35) we deduce the metric, the pressure and the density as a function of the still unknown quantity R(t).

10.4 The Robertson-Walker Metric

The metric element ds^2 (10.33) that we have arrived at is the *Robertson-Walker metric*, named after its two discoverers, or more briefly the *RW metric*. It lies at the foundation of model relativistic universes and is of fundamental importance in cosmology. Therefore, though we still have not determined Einstein's equations entirely, since R(t) is still unknown, we concentrate in this Section 4 on a study of the RW metric.

1. The RW metric (or element) is a very particular solution of Einstein's equations; this is apparent all through the derivation in the preceding paragraph, where we made many changes of variables that were arbitrary and introduced functions with the sole justification of getting a solution.

2. For the time variable t fixed, and thus dt = 0, the ds^2 obtained is that of a 3-space with constant Riemannian curvature at every point of the space. It suffices to refer to ds^2 in the conformal representation (9.11) to see that the Riemannian curvature is

$$K = k/R(t)^2.$$

For fixed t, the space is therefore non-Euclidean, having as its radius of curvature $R(t) = 1/\sqrt{|K|}$; this is respectively spherical, Euclidean or hyperbolic for k = 1, 0 or -1.

3. The assumption of isotropy with respect to the Earth leads to a space of constant curvature and, consequently, to a space that is isotropic in regard to all other points. This fact is remarkable enough, but even so we must not forget that isotropy with regard to the Earth is only observable within certain limits: redshifts up to 0.5 for galaxies, 5 for radio sources, 7 for the 3 K microwave background, and that anyway these observations are limited by the cosmological horizon. We shall touch on these issues later.

We express these properties of constant curvature by saying that the RW metric leads to *homogeneous model universes*.

4. For t fixed, the density ρ and pressure p each have the same values at every point in the space as shown by equations (10.34) and (10.35). This may be expressed by saying that we are dealing with *uniform model universes*.

5. The worldlines r = constant, θ = constant, φ = constant are geodesics of the spacetime. To verify that this is indeed so, merely replace in the geodesic equations (8.6) the g_{ij} by their values in (10.33).

This apparently innocuous result is actually fundamental. For on this rests the concept of an expanding universe. By virtue of the principle of geodesics (Section 1) the result signifies that the worldlines in question are possible solutions of the equations of General Relativity and they therefore represent possible motions. They can be identified with the motions of field galaxies (after correcting for random motions of around 200 - 300 km s^{-1}) or galaxy clusters. In these circumstances the field galaxies or clusters have fixed coordinates, r, θ, φ. These coordinates, in some sense locked into these objects, are termed *comoving coordinates*.

6. Then R(t) appears as a scaling parameter for the universe in (10.33). Effectively, for t fixed, the distance s between two points $\underline{M}_1$ and $\underline{M}_2$ with comoving coordinates r_1, θ_1, φ_1 and r_2, θ_2, φ_2 is given by the length of the geodesic arc $\underline{M}_1\underline{M}_2$ in the space, a length that is obtained on integrating ds along the arc; R(t) is a multiplying factor in this integration. Therefore, the distance between points $\underline{M}_1$, $\underline{M}_2$ is proportional to R(t). We note that r, like θ and φ, is dimensionless.

In particular, if R(t) increases with t, all distances in the space increase and we are led therefore to an expansion of the universe. Because it is R(t) that increases, we can say that the expansion of the universe is a phenomenon in which the space is itself increasing its dimensions, whereas the galaxies stay locked in fixed positions. The expansion is not a phenomenon in which the spatial structure remains fixed and the galaxies are moving; moreover, should the space be spherical, such a motion of galaxies would be impossible because they would be dispersing at one place and concentrating at some other place.

Often one encounters the notion that in this scheme a measuring rod would also expand and that, consequently, the distance between two galaxies, as measured with it, would be constant. This is not so because not each of the various atoms in the rod, governed by atomic interactions, would follow a geodesic in spacetime r = constant, θ = constant, φ = constant. The central atom could, for example, follow such a geodesic, but those at the extremities would not follow it. Consequently, some of the r, θ, φ coordinates would be varying with time in such a way that ds^2 for the rod remained constant.

7. The time t which appears in the RW line element is a very special time. It is a result of the very particular manner in which we derived the solution. But it has profound physical significance: at every point fixed in the comoving coordinates it is the *proper time*, that is to say the time that would be shown by any clock mechanism that did not have any reason to vary; tuning-fork, spring-watch, atom, radioactive nucleus, all unperturbed. Effectively, t is directly proportional to s for $dr = d\theta = d\varphi = 0$.

Furthermore, we can show that this time t is the proper time for various points fixed in comoving coordinates, synchronised by exchanges of signals propagating at the velocity c. It would be shown by clocks in various galaxies, synchronised by exchanging light signals. This time t is called *cosmic time*.

It is proper to remark here that this is not an absolute time; General Relativity is incompatible with such a concept. It owes its universal appearance to the fact that the universe is populated with objects whose comoving coordinates are fixed. If this were not so, if, for example, the galaxies in the northern hemisphere were to have the usual expansion but those in the south were to have a different one, solution (10.33) would not have its remarkable properties, and time t would not have the properties of cosmic time.

8. The comoving coordinates r, θ, φ are not an absolute system, either. They owe their particular properties to the fact that, effectively, the distribution of mass in the universe may be described by masses having fixed comoving coordinates. But, to give a counter-example, if the universe were populated, in addition to galaxies, by a comparable density of black holes all moving in a same direction (if that makes sense) with a significant velocity, then solution (10.33) would not be obtained with its characteristic properties.

These varied properties of the RW line element are fundamental and a cause for deep reflexion.

10.5 The Friedmann Universes

Everything we have set out so far is valid for a radius of curvature of the universe R(t) that varies in any fashion as a function of cosmic time t. To solve the cosmological problem completely it is necessary to determine the way in which R(t) effectively varies in our univérse.

At the end of 10.3 we had obtained two equations, the two Einstein equations (10.34) and (10.35) for determining the three functions R(t), ρ(t) and p(t). A third equation for solving the problem is therefore missing. This is given by the equation of state for the perfect fluid filling the space

$$p = p(\rho) \tag{10.36}$$

relating pressure p to density ρ.

A particularly simple equation of state is

$$p = 0. \tag{10.37}$$

This equation, introduced by Friedmann, is a very good approximation, for the galaxies that constitute the gas filling the universe have such small random velocities that their kinetic energy is very much less than their restmass energy. Furthermore, the energy of the radiation filling the universe is negligible and the perfect fluid behaves as a fluid of dust, whence the name *dust universe*, sometimes given to models based on equation (10.37).

In this paragraph we consider only the case in which the cosmological constant Λ is zero, returning in Chapter 12 to the case $\Lambda \neq 0$. The models which are thus derived are the simplest relativistic models, the *Friedmann models*, and are adequate for a first shot at interpreting observations in the relativistic zone.

With p and Λ zero, Einstein's equations reduce to

$$\left. \begin{array}{l} \dfrac{2\ddot{R}}{R} + \dfrac{\dot{R}^2}{R^2} = -\dfrac{kc^2}{R^2} \\[2mm] \dfrac{\dot{R}^2}{R^2} - \dfrac{8\pi G\rho}{3} = -\dfrac{kc^2}{R^2} \end{array} \right\}. \qquad (10.38)$$

Friedmann's first integral - Putting $Q = \rho R^3$ and differentiating the second equation we get

$$2\dot{R}\ddot{R} = \frac{8\pi G}{3} \frac{\dot{Q}R - \dot{R}Q}{R^2}.$$

Subtracting the second Einstein equation from the first gives

$$2\frac{\ddot{R}}{R} = -\frac{8\pi G\rho}{3}.$$

Finally, dividing the preceding relation by this last one, one sees that $\dot{Q} = 0$, and thus

$$Q \equiv \rho R^3 = \text{constant}. \qquad (10.39)$$

This is the first Friedmann integral, expressing the fact that the density varies as R^{-3}. That is to say, the quantity of matter enclosed in each volume element in comoving coordinates - the *covolume* element - is invariant in the course of evolution. There is neither creation nor destruction of matter in these models. The *codensity* is constant.

Friedmann's solution - Thanks to this first integral the system of differential equations (10.38) is simplified and leads to a simple first-order differential equation for R(t)

$$\dot{R} = \sqrt{8\pi GQ/3R - kc^2}. \qquad (10.40)$$

For k = 0 one solution is

$$R = (6\pi GQ)^{1/3} t^{2/3} \qquad (10.41)$$

where we have taken t = 0 when R is zero. This is the simplest solution, called the

Einstein-de Sitter model. It corresponds to a Euclidean space, in which it is not proper to speak of R as the radius of curvature - which is infinite - but as a simple distance scale factor giving the dimensions of every part of space throughout cosmic time. R is by definition a positive quantity (Section 4.2) so the dimensions of the universe decrease during epochs of negative cosmic time, reaching zero at t = 0, then increasing anew for t positive (Fig. 10.2).

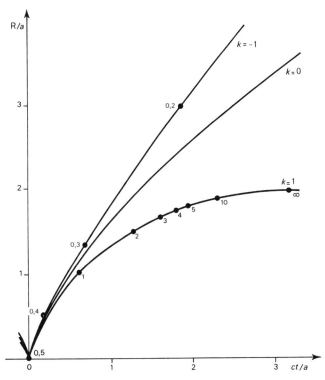

Fig. 10.2. Variation of R(t) in the course of cosmic time

For k = ± 1, one introduces a *scaling parameter*

$$a = 4\pi GQ/3c^2, \tag{10.42}$$

which allows the differential equation (10.40) to be written dimensionlessly

$$d(R/a)/\sqrt{2a/R - k} = d(ct/a).$$

Parametric solutions are obtained as a function of an angle δ termed the *development angle of the universe*:

$$\left.\begin{array}{ll} R/a = 1 - \cos \delta \\ ct/a = \delta - \sin \delta \end{array} \right\} \text{for } k = 1 \\ \left.\begin{array}{ll} R/a = \cosh \delta - 1 \\ ct/a = \sinh \delta - \delta \end{array} \right\} \text{for } k = -1 \right\}. \quad (10.43)$$

For $k = 1$, spherical space has a curvature radius varying in the course of time along a cycloid while for $k = -1$ the radius of hyperbolic space decreases from infinity to zero and then increases once more to infinity. These variations are shown in Fig. 10.2 which thus gives the geometrical history of the universe. As for the physical history, that is given by $p = 0$ and $\rho = Q/R(t)^3$.

The Friedmann model universes show that space can experience states of infinite density similar to those occurring during the gravitational collapse of a spherical mass according to the theory of General Relativity. This condensed state is unique in the hyperbolic and Euclidean cases but in the spherical case repeats regularly with period

$$P = 8\pi^2 GQ/3c^3. \quad (10.44)$$

We note, however, that the approximation $p = 0$ is not valid if the universe is sufficiently condensed, and we must consider more realistic equations of state to describe the condensed phase of the universe.

Here we make an interesting remark: during the evolution the value of k remains constant, and therefore the space retains its spherical, Euclidean or hyperbolic nature.

Observers prefer to use two other constants of integration: the *Hubble constant* and the *deceleration parameter* defined by

$$H = \dot{R}/R \quad (10.45)$$

$$q = -\ddot{R}R/\dot{R}^2. \quad (10.46)$$

The Hubble constant measures the rate at which the dimensions of the universe change, whereas the deceleration parameter measures the rate of change of this variation.

It can be shown that relations exist between the actual values H_0 and q_0 of H and q and the values a and δ_0:

$$a = \frac{c}{H_0} \frac{q_0}{|2q_0 - 1|^{3/2}} \quad (10.47)$$

$$\delta_0 = \left. \begin{array}{l} \cos^{-1} \\ \cosh^{-1} \end{array} \right\} \frac{1 - q_0}{q_0} \text{ for } k = \left\{ \begin{array}{l} 1 \\ -1 \end{array} \right. . \quad (10.48)$$

Note that the Hubble "constant" varies with time.

Various values of q are marked on the curves of Fig. 10.2. If space is hyperbolic, $q < \frac{1}{2}$ and if spherical, $q > \frac{1}{2}$, whereas for Euclidean space $q = \frac{1}{2}$. This explains the relentless enthusiasm with which observers try to determine the actual value of the deceleration parameter, a value on which the nature of the space depends.

We also have the following relations for the radius and density of the universe:

$$R = (c/H)/|2q - 1|^{1/2} \tag{10.49}$$

$$\rho = 3H^2 q/4\pi G. \tag{10.50}$$

The actual value of the Hubble constant, as we have already seen in the first part of this book, is $H_0 \approx 1/(12.2 \cdot 10^9 \text{ yr})$. Hence, the length c/H_0, which features in the formulae above, has the value $12.2 \cdot 10^9$ light years:

$$c/H_0 \approx 3.7 \text{ Gpc} \approx 1.2 \cdot 10^{28} \text{ cm}. \tag{10.51}$$

This shows the importance of H_0, on which the dimensions of the universe partly depend. The formula (10.50) also shows the importance of the density of the universe in cosmological problems; on its actual value, H_0 being more or less known, depends the value of q_0 and therefore the nature of space. For $q_0 = \frac{1}{2}$, that is to say for the Einstein-de Sitter cosmology, the value of the density is

$$\rho_{EdS} = 3H^2/8\pi G. \tag{10.52}$$

Often one uses the *density parameter* $\Omega = \rho/\rho_{EdS}$.

With the value of H_0 above we get

$$\rho_{EdS,0} = 1.2 \cdot 10^{-29} \text{ g cm}^{-3}; \tag{10.53}$$

above this value space is spherical, below it is hyperbolic. The observed value of the density of the universe indicates (as COUDERC first remarked) that the deceleration parameter is small. However, these conclusions refer to the case $\Lambda = 0$.

10.6 Radiation-Filled Universes

Up to what value for the density of the universe is the Friedmann model equation of state $p = 0$ valid? If the radius of the universe were 100 times smaller the galaxies would touch each other; if it were 10^7 times smaller the stars would touch, and for 10^{12} times the atomic nuclei would be in contact. For densities attaining such values it is probable, as we shall see in Chapter 17, that the universe would mainly be filled with intense radiation on account of the very high temperature resulting from this extreme compression.

The equation of state of a photon gas

$$p = \rho c^2/3 \tag{10.54}$$

should therefore be used for the epochs of great density.
With $\Lambda = 0$, Einstein's equations lead to

$$\rho R^4 \equiv Q' = \text{constant} \tag{10.55}$$

$$\dot{R} = \sqrt{8\pi G Q'/3R^2 - kc^2} \tag{10.56}$$

instead of the first integral (10.39) and differential equation (10.40).
SANDAGE has given the solution; thus in the Euclidean case we obtain

$$R = R_0(t/t_0)^{1/2}, \quad q \equiv 1, \tag{10.57}$$

where R_0 is the actual value of R.

There is not a great deal of difference between this and the dusty universe where one has $R \propto t^{2/3}$. In particular, the state of infinite density still exists and indeed appears to be an inherent characteristic of General Relativity which we could only overcome, perhaps, by the introduction of quantum phenomena. We shall say a few words about this in the final chapter.

Solutions have been given by CHERNIN for the case in which matter and radiation are simultaneously present, which leads progressively to a merging of the solutions of (10.56) and (10.40). HARRISON has similarly treated the case where $\Lambda \neq 0$.

By Stefan's law the temperature T of the photon gas is related to the density via

$$\rho = a_s T^4/c^2 \tag{10.58}$$

where a_s is Stefan's constant.

The first integral (10.55) gives

$$T = T_0 R_0/R; \tag{10.59}$$

the temperature of the radiation is inversely proportional to the radius of the universe.

11. Theory of Observations in the Relativistic Zone

General Relativity, with the hypothesis of a universe filled with a perfect fluid of density ρ and pressure p and a space that is isotropic with respect to the Earth, furnishes us with a cosmological solution whereby spacetime acts like a space with a radius of curvature $R(t)$, the same everywhere but varying in the course of cosmic time t; the ds^2 of Robertson-Walker, ρ and p are given as a function of $R(t)$, $R(t)$ depending finally on the equation of state $p = p(\rho)$ for a perfect fluid. In this chapter we give the basics governing the theory of observations made in the relativistic zone, for an arbitrary function $R(t)$. We apply the results particularly to the most interesting examples of the Friedmann universes - corresponding to $p = 0$ (and $\Lambda = 0$) - and, especially, to the Einstein-de Sitter model - corresponding in addition to the Euclidean case. Of course, this theory gives the usual classical non-relativistic Euclidean solutions for observations in the local neighbourhood.

11.1 Motion of Photons

Observations in the relativistic zone are made via photons: visible, radio, X, γ... It is therefore necessary at a fundamental level to know how these photons propagate from the object observed to us; how they eventually evolve along their trajectory in a curved universe with variable geometry. This is the first problem to resolve in the theory of the observations.

With the RW metric (10.33) let us consider a photon emitted by a galaxy G_e with comoving, fixed coordinates r_e, θ_e, φ_e, at a time t_e, observed by us in the Galaxy G_o, with comoving, fixed coordinates $r = 0$ at time t_o. This photon follows a geodesic curve of zero length in spacetime from G_e to G_o, a line that we may represent by parametric equations as a function of time

$r(t)$, $\theta(t)$, $\varphi(t)$.

We may show that $\theta(t) \equiv \theta_e$, $\varphi(t) \equiv \varphi_e$, that is to say that if we mark out the successive points in space through which the photon passes, we obtain a radial geodesic in the space: all the points thus marked are situated in the same direction θ_e, φ_e. This is entirely reasonable for reasons of symmetry, for space is isotropic and its expansion - or contraction - also. But, we can check it by starting from the differential equations of the geodesics (8.6) and using orthogonal coordinates to simplify the calculations.

This property is very important, for it leads to $d\theta = d\varphi = 0$ for the photon path, and thence, since $ds^2 = 0$, to the *equation for the motion of the photon*

$$c^2 dt^2 - R(t)^2 \frac{dr^2}{\left(1 + \frac{kr^2}{4}\right)^2} = 0 \tag{11.1}$$

whose solution $r(t)$ gives the motion of the photon. It is easiest to work with the reduced sinusoidal representaion (9.18). The integration leads to

$$c\int_{t_e}^{t_o} \frac{dt}{R(t)} = \int_0^{r_e} \frac{dr}{\sqrt{1-kr^2}} = \begin{cases} \sin^{-1} r_e & \text{if } k=1 \\ r_e & 0 \\ \sinh^{-1} r_e & -1 \end{cases} \tag{11.2}$$

where r_e, t_e, t_o correspond to the ds^2 (9.18).

Application to Friedmann models - Let us take for $R(t)$ the function (10.43) and make the first integration of (11.2) to obtain

$$\begin{aligned} r_e &= \sin(\delta_o - \delta_e) \quad \text{if } k = 1 \\ &= \sinh(\delta_o - \delta_e) \quad \text{if } k = -1. \end{aligned} \tag{11.3}$$

r_e is very simply tied to the angle of development. Via the second equations (10.43), we relate δ to t, and therefore r to t.

For the Einstein-de Sitter case, the solution is given still more simply and directly by (10.41):

$$r_e = (3ct_o/R_o)[1 - (t_e/t_o)^{1/3}]. \tag{11.4}$$

11.2 Spectral Ratio

Having determined the movement of a photon, we shall now study its evolution. Let us consider, always in the reduced sinusoidal representation, a signal emitted from r_e at t_e, and another emitted from r_e at $t_e + dt_e$. These two successive signals arrive at $r = 0$ at t_o and $t_o + dt_o$. Equation (11.2) gives

$$c\int_{t_e}^{t_o} \ldots = \int_0^{r_e} \ldots$$

and

$$c\int_{t_e + dt_e}^{t_o + dt_o} \ldots = c\int_{t_e + dt_e}^{t_e} \ldots + c\int_{t_e}^{t_o} \ldots + c\int_{t_o}^{t_o + dt_o} \ldots = \int_0^{r_e} \ldots$$

therefore

$$\int_{t_0}^{t_0 + dt_0} \ldots = \int_{t_e}^{t_e + dt_e} \ldots$$

or

$$dt_0/R_0 = dt_e/R_e.$$

In particular, let us suppose that dt_e corresponds to the time interval between two successive maxima of an electromagnetic wave emitted at r_e: the periods of emission P_e and receipt P_0 are then related by

$$P_0/R_0 = P_e/R_e,$$

and, since the velocity of light is constant, the wavelengths obey the relation

$$\lambda_0/R_0 = \lambda_e/R_e. \tag{11.5}$$

(11.5) is a fundamental relation in observational cosmology. It states that the wavelength of a photon is proportional, in the course of its propagation, to the radius of curvature of the space or to its scale, this wavelength being measured by observers in fixed comoving coordinates. This last condition is important: if an observer has variable comoving coordinates, an ordinary Doppler effect is superimposed on the wavelength variations implied by (11.5).

The photon considered undergoes a *spectral shift* ("*redshift*") z given by

$$z \equiv (\lambda_0 - \lambda_e)/\lambda_e = R_0/R_e - 1. \tag{11.6}$$

But it seems to us more natural, in view of relation (11.5), and moreover it simplifies the equations we shall consider, to introduce a quantity ζ which we shall call the *spectral ratio*:

$$\zeta \equiv \lambda_0/\lambda_e = R_0/R_e = 1 + z. \tag{11.7}$$

If $R_e < R_0$, we have a true redshift (for visible photons), and this whatever the behaviour of $R(t)$ between times t_e and t_0 is. If $R_e > R_0$, we have a blueshift.

Historically the concept of distance has been associated with redshift in the following manner: for a nearby galaxy

$$z \approx \frac{R_0}{R_0 - \dot{R}_0 \Delta t} - 1 \approx \frac{\dot{R}_0}{R_0} \Delta t$$

where $\Delta t = t_0 - t_e$ is small. But $\Delta t = D/c$, D being the distance of the observed galaxy, which here is defined without ambiguity. If the redshift is interpreted as a Doppler effect, we have $z = v/c$, where v is the radial velocity of the galaxy, and therefore $v \approx (\dot{R}_0/R_0)D$, which is precisely the Hubble law according to which velocity is proportional to distance, with a coefficient of proportionality $H = \dot{R}_o/R_o$.

But, in fact, Hubble's law thus expressed is only a secondary consequence, valid to first order, of the fundamental law of General Relativistic cosmology according to which wavelengths are proportional to the radius of curvature of the space or to its scale.

In the relativistic zone of observation, use of the notion of distance is to be avoided because it does not have a really precise meaning: is it concerned with distance at the instant of emission, distance at the moment of observation, or at what other moment? Is distance to be measured using the apparent diameter, or the apparent brightness, for which, as we are going to see, surprises await us?

The fundamental physical quantity, directly measurable without ambiguity, is the spectral ratio ζ - or the spectral shift z. We shall express the main observables for an object as a function of this, using it in place of the "distance", with which there is a simple, unique, and unequivocal correspondence only for the nearby zone of space. At most we could use the comoving radial coordinate r_e as a distance. If R(t) is a single-valued function (in practice this is generally the case) we go from ζ to r_e through the chain

$$\zeta \underset{(11.7)}{\leftrightarrow} R_e \underset{R(t)}{\leftrightarrow} t_e \underset{(11.2)}{\leftrightarrow} r_e. \qquad (11.8a)$$

For the Einstein-de Sitter case this chain reduces to

$$r_e = \frac{3ct_o}{R_o}(1 - \zeta^{-1/2}). \qquad (11.8b)$$

But r_e is a dimensionless radial coordinate; furthermore, its value depends primarily on the representation chosen for ds^2.

Thus, we shall use ζ in the following to derive quantities that interest us, being free to use subsequent elimination of ζ in order to find other relations between the quantities.

11.3 Travel Time of Photons

The first quantity that naturally presents itself is the interval of cosmic time τ which elapses between the emission from an object and our observation of the photon. This interval τ is very important because it gives us the age of the received image of the object and the instant t_e at which we are seeing it. We have to work this out with care because, in the relativistic zone, it gives us information about phenomena that have already taken place in the remote history of the universe.

The calculation is readily made since

$$\tau = t_o - t_e. \qquad (11.9)$$

Application to Friedmann models - Equations (11.3) give t from R. So for k = 1, we get two relations

$$\tau = t_0 \left(1 - \frac{\delta_e - \sin \delta_e}{\delta_0 - \sin \delta_0}\right), \quad \zeta = \frac{1 - \cos \delta_0}{1 - \cos \delta_e},$$

the second of which gives δ_e as a function of ζ and δ_0 which, substituted in the first, gives τ as a function of ζ, t_0 and δ_0, in other words of ζ and actual properties of the universe, properties that can be expressed by H_0 and q_0. These calculations have been made numerically by SANDAGE, as shown in Fig. 11.1 which reproduces the result for several values of q_0.

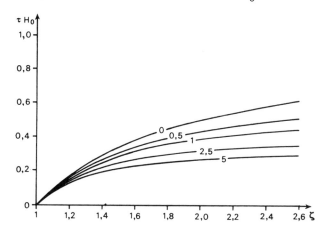

Fig. 11.1. Travel time of photons

In the Einstein-de Sitter case, the calculations are much simpler. They directly give $t_0/t_e = \zeta^{3/2}$, whence, as $H_0 = \dot{R}_0/R_0 = 2/3t_0$:

$$\tau_{EdS} = 2(1 - \zeta^{-3/2})/3H_0. \tag{11.10}$$

Compared to the classical result $\tau_{Cl} = z/H_0$, obtained at the end of Section 11.2, (11.10) shows the enormous error that would be made with this latter result for large values of ζ. For quasars at $\zeta = 2$, the error is a factor of 2.

We can also see that for $\zeta = \infty$, i.e. for an object at the observational limit, as the photons reach us with vanishingly-small energy, τ_{EdS} attains a limit

$$\tau_{lim} = 2/3H_0. \tag{11.11}$$

It is impossible to see anything further back in time than τ_{lim} which, as we shall see, is just the elapsed time since the epoch when R(t) was zero. Right to the limits of large radio telescopes, where ζ approaches 9, one can observe objects in the state in which they were only 250 million years after the beginning of the expansion. That it is possible to go back in this way through the history of the universe is quite fantastic.

11.4 Age of the Universe

We can define the age of the universe as the time t_0 that has elapsed since its era of very high density.

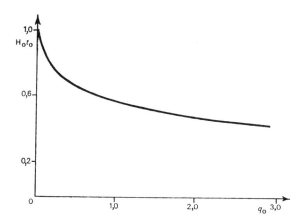

Fig. 11.2. Age of the universe

To obtain this it is sufficient to set $t_e = 0$ in equation (11.9). For the Friedmann models, this calculation also is due to SANDAGE: the result is given graphically in Fig. 11.2, as a function of H_0 and q_0. For a given value of H_0 the largest value of t_0 is given for the hyperbolic model with $q_0 = 0$. The age of the universe is then $1/H_0 = 12.2 \cdot 10^9$ yr for $H_0 = 80$ km s^{-1} Mpc^{-1}. In the Einstein-de Sitter case it would have 2/3 this value, being only $8.1 \cdot 10^9$ yr.

11.5 Diameter

In static Euclidean space the linear diameter of an object is readily deduced from its distance and apparent diameter. What happens in an expanding space where the concept of distance loses precise meaning and, furthermore, when space is curved?

To elucidate this question consider two photons emitted from two points $\underline{A}$, $\underline{B}$ with comoving coordinates r_e, 0, 0 and r_e, θ, 0 at time t_e, θ being small. These photons reach us at t_0, and since their trajectories are radial, they arrive along directions separated by the angle θ, which is therefore the apparent angular diameter of object $\underline{AB}$. On the other hand, the linear separation Θ of two points $\underline{AB}$, at the instant of emission t_e, is given by ds^2 in the reduced sinusoidal representation

$$ds^2 = 0 - R_e^2 \left(\frac{0}{1 - kr_e^2} + r_e^2 \theta^2 + 0 \right) = -\Theta^2$$

whence

$$\theta = \Theta/R_e r_e \tag{11.12}$$

which, working through chain (11.18), lets us express the apparent diameter θ as a function of the linear diameter Θ and the spectral ratio ζ.

Application to Friedmann models - For the Friedmann models the quantity $R_e r_e$ in the reduced sinusoidal representation has been calculated by MATTIG as a function of ζ, H_0 and q_0:

$$R_e r_e/(c/H_0) = \frac{q_0 \zeta + 1 - 2q_0 + (q_0 - 1)\sqrt{2q_0 \zeta + 1 - 2q_0}}{q_0^2 \zeta^2}. \tag{11.13}$$

Two particular useful and simpler cases are

$$R_e r_e/(c/H_0) = (\zeta^2 - 1)/2\zeta^2 \quad \text{for } q_0 = 0 \tag{11.14}$$
$$= (\zeta - 1)/\zeta^2 \quad \text{for } q_0 = 1. \tag{11.15}$$

In the Einstein-de Sitter case $q_0 = \frac{1}{2}$, and we have

$$R_e r_e/(c/H_0) = 2(1 - \zeta^{-1/2})/\zeta, \tag{11.16}$$

whence

$$\theta_{EdS} = \frac{\Theta}{c/H_0} \frac{\zeta}{2(1 - \zeta^{-1/2})}. \tag{11.17}$$

The quantity $R_e r_e/(c/H_0)$, or $R_0 r_e/(c/H_0)\zeta$, enters numerous calculations. It has been tabulated by SANDAGE (who names it A) and calculated by McVITTIE for certain cases when the cosmological constant Λ is non-zero (he calls it $\mathcal{D}$).
Figure 11.3, due to SANDAGE, gives the base 10 logarithm for the apparent diameter θ thus obtained, in arbitrary units, as a function of the spectral ratio ζ,

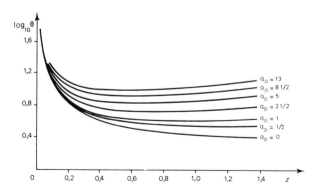

Fig. 11.3. Apparent diameter of an object

for various values of q_0. Compared to the classical result

$$\theta_{Cl} = \frac{\Theta}{c/H_0} \cdot \frac{1}{\zeta - 1} \tag{11.18}$$

this figure shows the very bizarre behaviour of apparent diameters. For small redshifts it accords with the classical result, varying like $1/z$, but as z becomes large it decreases less quickly. In the Einstein-de Sitter case, we get around three times the classical diameter for $z = 1$. It even reaches a minimum value for $z = 5/4$ and begins to increase for yet larger redshifts. An analogous behaviour takes place in spherical space, the apparent diameter of an object passing through a minimum as the object reaches the equatorial surface of the observing point (see Section 9.2). But here a similar effect does not come into play, since space is Euclidean; in fact, the increase of the apparent diameter for large z is due to the fact that we pick up photons which were emitted when the object was near us because the universe was more compact. Whimsically, we could say that the further we look the nearer we see.

11.6 Luminosity

A fundamental quantity we need is the absolute magnitude or radiant energy of an object. We give the principles of calculating it as a function of apparent magnitude, or the energy flux received, and the spectral ratio.

We compute the number Δn of photons emitted by the source during a brief time Δt_e. We calculate the area S_0 - using non-Euclidean formulas (9.26) and (9.24) - of the sphere where the photons are located at the moment of observation t_0, as well as the time Δt_0, also small, during which these photons sweep past the observer. We calculate the change in energy of the photons on arrival, due to the redshift.

The area of the sphere at t_0 is equal to that of a sphere centred on us at $r = 0$ and with comoving radial coordinate $r = r_e$; on this sphere we have $ds^2 = R_0^2 r_e^2 da^2$, therefore its area is

$$S_0 = R_0^2 r_e^2 \int da^2 = 4\pi R_0^2 r_e^2.$$

Furthermore, we have $\Delta t_0 = \zeta \Delta t_e$, and the energy of a received photon is $hc/\zeta\lambda_e$.

The received *energy flux*, that is to say the energy received per unit time and per unit area is

$$f_0 = \Delta n \frac{hc}{\zeta\lambda_e} \frac{1}{\zeta\Delta t_e} \frac{1}{4\pi R_0^2 r_e^2} = P_e/\zeta^2 R_0^2 r_e^2 \tag{11.19}$$

where P_e is the power radiated per steradian at emission.

For Friedmann models, $R_0 r_e$ is calculable directly from the formula of MATTIG (11.13), since we have $R_0 r_e = \zeta R_e r_e$.

In the Einstein-de Sitter case, we obtain for the radiant power

$$P_e = \zeta^2 (1 - \zeta^{-1/2})^2 \, 4c^2 f_0/H_0^2. \tag{11.20}$$

Apparent bolometric magnitude - The *apparent bolometric magnitude* is

$$m_{bol} = -2.5 \log_{10} f_0 + \text{constant} \tag{11.21}$$

where the constant is expressed as a function of the absolute magnitude (see (11.24)). For the Einstein-de Sitter universe we have

$$m_{bol.EdS} = 5 \log_{10} (\zeta - \zeta^{1/2}) + \text{constant}, \tag{11.22}$$

whereas the classical formula gives

$$m_{bol.Cl} = 5 \log_{10} \frac{\zeta - 1}{2} + \text{constant}. \tag{11.23}$$

There is a small difference between $m_{bol.EdS}$ and $m_{bol.Cl}$: only one magnitude for $\zeta = 10$. The flux behaves essentially like $1/z^2$ in the Einstein-de Sitter case, just like in the classical case. For $q_0 = 1$ it behaves exactly like $1/z^2$. Figure 11.4 gives, according to SANDAGE's calculations, m_{bol} as a function of z for various values of q_0, in the case of the brightest galaxies in clusters.

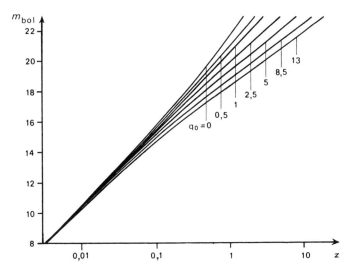

Fig. 11.4. Apparent bolometric magnitude of a galaxy

The fact that m_{bol} is quite insensitive to the value of the deceleration parameter ruins the method of finding q_0 from the magnitude-redshift relation for galaxies.

Distance modulus - The absolute bolometric magnitude M_{bol} is the apparent bolometric magnitude for an object situated at $d_0 = 10$ pc from us. We obtain it therefore from (11.21) on putting $R_0 r_e = d_0$ in (11.19). The *distance modulus* is the difference

$$\mu \equiv m_{bol} - M_{bol} = 5 \log_{10} \frac{\zeta R_0 r_e}{c/H_0} + 5 \log_{10} \frac{c/H_0}{d_0}. \tag{11.24}$$

With $H_0 = 80$ km s^{-1} Mpc^{-1}, the last term is 42.9. The first term is calculated by MATTIG's formula for the Friedmann models. For example, we obtain for the Einstein-de Sitter model

$$\mu_{EdS} = 5 \log_{10} (\zeta - \sqrt{\zeta}) + 44.4 \tag{11.25}$$

which agrees with the classical formula for ζ approaching 1:

$$\mu_{Cl} = 5 \log_{10} (\zeta - 1) + 42.9; \tag{11.26}$$

for $\zeta = 10$, μ_{EdS} and μ_{Cl} differ by only one magnitude.

Flux density - Thus far we have considered the energy emitted or received independently of wavelength, for we were interested only in bolometric luminosities. But in general, and above all in radio astronomy, the receivers are highly selective, receiving only the energy in a chosen frequency band and measuring therefore a *flux density*, that is to say the energy received per unit time, per unit area, per unit bandwidth.

Let $F(\nu_e)$ be the emission spectrum of an object, that is to say the power radiated per unit solid angle per unit bandwidth. The flux density received is calculated according to the following scheme:

- we receive in a band $(\nu_0, \nu_0 + d\nu_0)$;
- at emission the frequency is $\nu_e = \zeta \nu_0$;
- the energy emitted in time Δt_e in a band $d\nu_e$ is $4\pi F(\nu_e) d\nu_e \Delta t_e$;
- the number of photons is obtained by dividing this quantity by $h\nu_e$;
- these photons, on receipt, are spread over an area $4\pi R_0^2 r_e^2$ and along a time $\Delta t_0 = \zeta \Delta t_e$;
- the received energy is therefore $4\pi F(\nu_e) d\nu_e \Delta t_e \frac{h\nu_0}{h\nu_e}$;
- at last the flux density received is

$$s_0 = \frac{4\pi F(\nu_e) d\nu_e \Delta t_e \nu_0}{\nu_e 4\pi R_0^2 r_e^2 \Delta t_0 d\nu_0} = F(\zeta \nu_0)/\zeta R_0^2 r_e^2. \tag{11.27}$$

MATTIG's formula (11.13) also permits us to calculate s_0 as a function of ζ, H_0 and q_0. For the Einstein-de Sitter case we have

$$s_{0.EdS} = F(\zeta\nu_0)/(c/H_0)^2 \, 4\zeta(1-\zeta^{-1/2})^2, \tag{11.28}$$

a very useful formula in radio astronomy for quasars and radio galaxies. In general, the spectrum follows a power law

$$F(\nu) = f\nu^{-\alpha} \tag{11.29}$$

where α is the *spectral index*. In this case we have

$$s_0 = f\nu_0^{-\alpha}/\zeta^{\alpha+1} \, R_0^2 \, r_e^2. \tag{11.30}$$

11.7 Brightness

For an object of area θ^2 and uniform brightness the apparent *bolometric brightness* is given by

$$B_{bol} = f_0/\theta^2 = P_e/\zeta^4 \theta^2 \tag{11.31}$$

and the apparent *monochromatic brightness* by

$$B_0 = s_0/\theta^2 = F(\zeta\nu_0)/\zeta^3 \theta^2 \tag{11.32}$$

with the same notation.

These results are independent of the model of the universe employed. But, however, the brightnesses decrease very rapidly if ζ increases, above all the bolometric magnitude, even though in the classical case it is constant. The apparent diameter of an object, measured as the apparent diameter out to a certain limiting isophote, fixed by the observer, will differ from that given by the relations in 11.5, unless the object has a sharp edge, as for example in the case where the intrinsic brightness is constant inside a certain contour and zero outside.

In the case where the spectrum follows a power law with spectral index α, the monochromatic brightness varies as $1/\zeta^{\alpha+3}$.

11.8 Number of Observable Objects

In a static Euclidean space uniformly populated with identical objects, the number of those that are observable up to a certain distance is proportional to the cube of that distance, and therefore to the total luminosity to the power -3/2, since luminosity is inversely proportional to the square of the distance.

In curved space with variable radius the law giving the number of observable objects out to a certain limiting spectral ratio - termed the *number count* - is very complicated, depending on the geometry of space and the form of its variation with cosmic time. This fact explains the enormous effort that astronomers have put into the determination of number counts, by observing galaxies to the optical limit, then quasars and radio galaxies in the optical and radio domains, in the hope of gleaning interesting cosmological data.

Let us consider a certain class of objects with a spatial concentration n, having the actual value n_0 - for example in galaxies Mpc^{-3} or quasars Gpc^{-3} - such that $n \propto R(t)^{-3}$, in other words such that the concentration in comoving coordinates, the *coconcentration* δ, would remain constant through cosmic time. We assume, stated otherwise, that the objects are permanent.

The covolume of space within the comoving radial coordinate r is $V(r)/R^3$, given by (9.22). Therefore the number of objects with radial coordinate less than r is $\delta V(r)/R^3$.

Expressing r as a function of ζ by the chain (11.8), the number of objects with spectral ratio less than ζ is

$$N(\zeta) = n_0 V[r(\zeta)] R_0^3/R^3. \qquad (11.33)$$

If R(t) is not single-valued, we must take good care to define what is meant by $N(\zeta)$. All the same, if space is spherical, we may include in the calculation the same object several times if its photons reach us after several circuits around space.

In the Einstein-de Sitter case we readily obtain

$$N_{EdS}(\zeta) = \frac{32}{3} \pi n_0 (c/H_0)^3 (1 - \zeta^{-1/2})^3. \qquad (11.34)$$

For an infinite spectral ratio, the limit of observation, the number of objects reaches a limit even though space is infinite:

$$N_{EdS.lim} = \frac{32}{3} \pi n_0 (c/H_0)^3. \qquad (11.35)$$

Figure 11.5 shows the result for an actual number density of 1 Mpc^{-3} and for H_0 = 80 km s^{-1} Mpc^{-1} and compares this to the classical result

$$N_{Cl} = \frac{4}{3} \pi n_0 (c/H_0)^3 (\zeta - 1)^3. \qquad (11.36)$$

We can only see 1800 billion objects. In contrast to a naive idea this number increases with time for the Einstein-de Sitter universe because H_0 diminishes: the expansion of the universe will not isolate us, but the opposite. In the race between light and expansion, the expansion is the less rapid. We shall come to this subject in more detail in the chapter on cosmological horizons.

The figure also shows the absolute necessity of using a relativistic formula in place of the classical formula: for ζ = 2 the classical formula gives a number that is already five times too large.

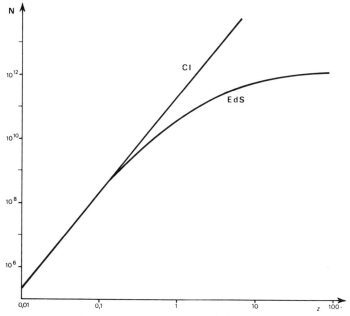

Fig. 11.5. The number of observable objects

In most astronomical investigations, number counts are made not as a function of spectral ratio but as a function of apparent magnitude or flux density, whichever is easier to measure. To obtain the number $N(m_{bol})$ of objects up to a certain apparent bolometric magnitude, all that is needed is to eliminate ζ between relations (11.33), (11.21) and (11.24). Tables have been given by SANDAGE for Friedmann models.

In the same manner, to have the number $N(s_o)$ of radio sources, having a certain emission spectrum $F(\nu)$, and observed to have a flux density exceeding a certain value s_o, it is enough to eliminate ζ between (11.33) and (11.27), or more particularly (11.30) if $F(\nu)$ has the form (11.29). Numerical calculations have been made by GLANFIELD for various values of spectral index and cosmological constant.

11.9 Other Parameters

Other parameters useful for the interpretation of observations are calculated along similar lines. Everything, finally, depends on judicious use of ds^2 and the formulas that follow from it. The ds^2 is the length standard or time standard to which it is always necessary to refer in order to pass from some coordinate system - most frequently the reduced sinusoidal representation - to real intrinsic values of length and elapsed time.

Thus we may calculate the number of objects of a certain category that a photon will encounter in its journey to us. This problem is important in questions of absorption and diffusion. Let us suppose that space is uniformly distributed with spherical objects having a linear diameter Θ that is independent of cosmic time, with an actual concentration n_o and a constant coconcentration. The likely number of such objects (atoms, galaxies...) that a photon emitted by an emitter at spectral ratio ζ will meet before reaching us is

$$N_p = \frac{\pi}{4} \Theta^2 n_o R_o^3 \int_0^{r(\zeta)} dr/R^2 \sqrt{1 - kr^2}. \tag{11.37}$$

In the Einstein-de Sitter model, from (11.4) and (11.7), we get

$$N_{p.EdS} = \frac{\pi}{6} \Theta^2 n_o (c/H_o) (\zeta^{3/2} - 1). \tag{11.38}$$

Similarly, if one wants to calculate, in the same notation, the mean brightness contributed to the celestial sphere by objects of a certain type - radio galaxies for example - up to a particular spectral ratio ζ, one calculates the number of objects contained in a spherical shell $(r, r+dr)$: $4\pi nR^3 r^2 dr/\sqrt{1-kr^2}$; from each we receive a flux density given by (11.30) for a powerlaw spectrum; the total flux density is

$$s_o(\zeta) = 4\pi n_o f \nu_o^{-\alpha} R_o \int_0^{r(\zeta)} dr/\zeta^{\alpha+1} \sqrt{1 - kr^2}$$

and the brightness is

$$B_o(\zeta) = s_o(\zeta)/4\pi. \tag{11.39}$$

In the Einstein-de Sitter case, (11.39) gives

$$B_{o.EdS}(\zeta) = n_o f \nu_o^{-\alpha} (c/H_o) (1 - \zeta^{-\alpha-3/2})/(\alpha + 3/2). \tag{11.40}$$

Here, too, for ζ infinite, the brightness reaches a limiting value

$$B_{o.EdS}(\infty) = n_o f \nu_o^{-\alpha} (c/H_o)/(\alpha + 3/2). \tag{11.41}$$

12. The Cosmological Constant

The cosmological constant forms a subject of debate among astronomers. In Einstein's equations (10.15) we must equate an energy-momentum tensor, symmetrical and of zero divergence, to a geometry tensor which also must be symmetrical and have zero divergence. One demonstrates, as indicated in Chapter 10, that the most general geometry tensor to fit those criteria is the tensor S^{ij} given by (10.14):

$$S^{ij} = R^{ij} - \frac{1}{2}g^{ij}(S - 2\Lambda),$$

where Λ, the cosmological constant, is arbitrary.

To be sure, S^{ij} without Λ also satisfies the conditions and for simplicity and the desire not to introduce into the theory a number of parameters that is too great, given the information supplied by the data, one may decide to introduce a form of S^{ij} without Λ into Einstein's equations. But in recent years the field of observational cosmology has been relatively enriched - but not enough to satisfy astronomers - and it is rather tightly constrained within the limits imposed by the absence of the cosmological constant.

Thus it is really legitimate, and moreover it does satisfy any worry we might have about generality, to introduce an S^{ij} with Λ into Einstein's equations and thence to explore the possibilities thus opened up. The observational data are readily accommodated, without being swamped in clothes a size too large.

In fact, without absolutely asserting that the cosmological constant must be introduced, we think that one may introduce it and that the situation is sufficiently ripe for the observations to be on the point of giving the answer that Λ does have some definite value. And furthermore that the value zero, sustained by no principle or theorem, is quite unlikely.

In this chapter we will show the forms of the variations of the radius of curvature of space R(t) versus cosmic time, for universes of zero pressure with $\Lambda \neq 0$ and we will see what effects the cosmological constant has on the interpretation of certain observations.

12.1 The (q,λ) Diagram

In Section 10.5 we considered the case when the equation of state for a perfect fluid filling the universe is p = 0 and where Λ = 0 (Friedmann models). Here we again take p = 0, but Λ has some value.

Einstein's equations (10.34) and (10.35) are then written

$$\left.\begin{aligned} \frac{2\ddot{R}}{R} + \frac{\dot{R}^2}{R^2} &= -\frac{kc^2}{R^2} + \Lambda \\ \frac{\dot{R}^2}{R^2} - \frac{8\pi G \rho}{3} &= -\frac{kc^2}{R^2} + \frac{\Lambda}{3} \end{aligned}\right\} \quad (12.1)$$

We introduce anew Hubble's constant H and the deceleration parameter q (10.45) and (10.46) and, furthermore, the *reduced cosmological constant* λ and the *density parameter* σ defined by

$$\lambda = \Lambda/3H^2 \qquad (12.2)$$

$$\sigma = 4\pi G \rho / 3H^2. \qquad (12.3)$$

q, λ and σ are dimensionless. λ varies with time via H, and $\sigma = \rho/2\rho_{EdS} = \Omega/2$ where ρ_{EdS} is the density (10.52) in the Einstein-de Sitter model.

With these parameters, equations (12.1) give

$$2q - 1 + 3\lambda = kc^2/H^2R^2 \qquad (12.4)$$

$$q + \lambda = \sigma. \qquad (12.5)$$

If we consider a plane (q,λ) where each point represents a possible state for the universe, the preceding relations permit us to introduce the boundaries of important regions.

- Universes of zero density are represented by the line q + λ = 0, and since the density of the universe is not negative, the area beneath the bisecting line and hatched in the (q,λ) diagram of Fig. 12.1 is forbidden.
- Euclidean spaces are represented by line E: 2q - 1 + 3λ = 0; above this line space is spherical and below it is hyperbolic.

The values estimated in the first part of this book for the density of the universe give a representative point in the (q,λ) diagram quite near the bisecting line unless the universe contains a great deal of hidden mass.

Notice that for Λ = 0, and therefore λ = 0, we recover Friedmann universes, on the q-axis, hyperbolic in the section (0,1/2), Euclidean at 1/2, and spherical in the portion (1/2, ∞). The Einstein-de Sitter universe is represented by the point q = 1/2, λ = 0.

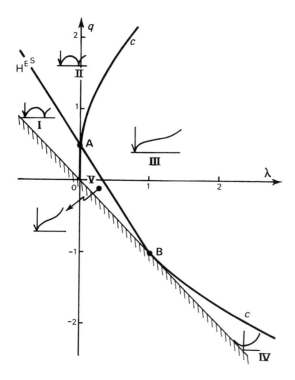

Fig. 12.1. The (q,λ) diagram

12.2 Evolution of the Universe

The solution of Einstein's equations (12.1) leads, as in the case $\Lambda = 0$, to the first Friedmann integral (10.39). Setting

$$y = R(t)/R(t_0), \tag{12.6}$$

Einstein's second equation becomes

$$\dot{y}^2 = H_0^2 \, \Phi(y) \tag{12.7}$$

where $\Phi(y)$ has the form $a/y + b + cy^2$:

$$\Phi(y) = \frac{2(q_0 + \lambda_0)}{y} + (1 - 2q_0 - 3\lambda_0) + \lambda_0 y^2, \tag{12.8}$$

the suffices o serving to indicate actual values.

The first order differential equation (12.7) has solutions given by the functions

$$H_0 t = \int dy / \sqrt{\Phi(y)} \tag{12.9}$$

which are elliptical functions on account of the form of $\Phi(y)$.

Following the theory of elliptical functions, there are three main cases to consider:

1) $b > 0$: then $k = -1$, space is hyperbolic. There are three sub-divisions:

$c < 0$: therefore $\Lambda < 0$; $R(t)$ oscillates from zero to a maximum, like a cycloid; this is sketched in Fig. 12.1 by the little graph at extreme left.

$c = 0$: therefore $\Lambda = 0$; $R(t)$ is monotonic for $t > 0$, growing from zero to infinity; asymptotically it increases like t. This is a Friedmann case already encountered.

$c > 0$: then $\Lambda > 0$; $R(t)$ again increases monotonically for $t > 0$, but it has a point of inflexion, shown in the small sketch at lower left in Fig. 12.1.

Hyperbolic oscillating spaces exist; we have seen in the second part that closed locally hyperbolic spaces exist. Therefore we make an important remark: the properties of oscillation or closure are not, as is often alleged, attributable only to spherical model universes.

2) $b = 0$: then $k = 0$, and the space is Euclidean. We obtain the same forms as above for each value of c, but simpler ones; $R(t)$ can be expressed using $\cos t$, $t^{2/3}$ (Einstein-de Sitter) or $\cosh t$.

3) $b < 0$: then $k = 1$, space is spherical. Five subdivisions are considered:

$c < 0$: $\Lambda < 0$, oscillating space, sketched by the uppermost graph in Fig. 12.1.

$c = 0$: $\Lambda = 0$; we have already met this oscillating Friedmann case.

$0 < c < c_{cr}$, where c_{cr} is a critical value such that equation $a + by + cy^3 = 0$ has a double root: $c_{cr} = -4b^3/27a^2$. To equation $c = c_{cr}$ there corresponds one third-degree curve on the (q,λ) plane, of which two useful portions are represented in Fig. 12.1 by the two arcs C.

In this sub-division we may therefore distinguish two possibilities: in the region above the upper arc C and to the right of the q-axis $R(t)$ again oscillates; in the region beneath the lower arc C, $R(t)$ passes through a minimum - sketched at lower right.

$c = c_{cr}$: we are on the arcs C in the (q,λ) diagram. In this case $R(t)$ has a horizontal asymptote (Fig. 12.2). On the upper arc C, $R(t)$ varies along curve E-E - these are the *Einstein-Eddington models*. On the lower arc C, $R(t)$ varies along the curve E-L - these are the *Eddington-Lemaître models*. Finally, at the point that is at infinity for the two arcs C, $R(t)$ is constant - this is the *Einstein model* E.

$c > c_{cr}$: $R(t)$ is again monotonic, for $t > 0$, increasing from zero to infinity with a point of inflexion - the central sketch in Fig. 12.1. These cases correspond to the *Lemaître models*, in which the radius of the universe (spherical) coasts for a long time in the neighbourhood of a constant value.

We remark that there exist non-oscillating models of spherical space.

Einstein's static space in one solution for which Einstein introduced the cosmological constant, a constant which he later rejected, because of his hope, prior to the discovery of the expanding universe, of obtaining an unchanging universe. This "historic" universe is truly one of the most particular solutions. The constancy of

R(t) gives $\dot{R} = 0$; equations (12.1) then give a particular value for the cosmological constant:

$$\Lambda_E = 4\pi G\rho. \qquad (12.10)$$

Therefore $\Lambda > 0$ and $k > 0$: the space is spherical and its radius of curvature is

$$R_E = c/2\sqrt{\pi G\rho}. \qquad (12.11)$$

With $\rho = 3 \cdot 10^{-31}$ g cm^{-3}, R_E is 19 Gpc; the circumference of the universe is 120 Gpc and its volume $3.6 \cdot 10^7$ Gpc^3. But this universe, "reassuring" by its closure and constancy, is not expanding.

Another very particular case corresponds to the point $\lambda = 1$, $q = -1$, the *de Sitter universe*. This is a Euclidean universe which is empty; its scale parameter varies as shown by curve E-L of Fig. 12.2 in which the asymptote E would have a null ordinate; actually, $R(t) \propto e^t$; this universe expands exponentially.

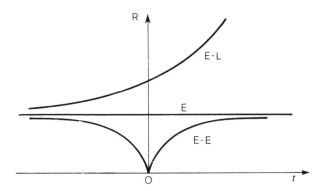

Fig. 12.2. The radii of curvature for the Eddington-Lemaitre, Einstein, and Einstein-Eddington models

The *Milne universe*, for which $R \propto t$, is also rather special; represented by the point $\lambda = 0$, $q = 0$, it is hyperbolic and empty.

Summarising, analysis of the solutions (12.9) of the differential equation (12.7) has led us to divide the (q,λ) diagram into five regions I,..., V, where space may be spherical or hyperbolic, with a null or positive minimum size, oscillating or monotonic (for $t > 0$).

All models with negative cosmological constant are oscillating. In this case, after an expansion phase, the universe contracts on itself, as if a negative value of Λ had the effect of stabilising it. For this reason the cosmological constant is sometimes interpreted as a *cosmic repulsion*.

If we knew the actual values of H, q and ρ with precision, we could locate the point representing our universe in the (q,λ) diagram and then deduce its nature and evolution. This evolution takes place along certain curves in the (q,λ) plane, obtained in parametric form from the functions $q(t)$ and $\lambda(t)$. These functions are calculated from $\ddot{R}/R$ and $\dot{R}/R$, once $R(t)$ is determined.

These evolutionary tracks have been numerically calculated by STABELL and REFSDAL. They pass through points A and B with coordinates (1/2, 0) and (-1,1), are tangential to the q-axis at A and to the line $\rho = 0$ at B, except for the degenerate cases of empty universes which stay on the line $\rho = 0$.

In regions I and II, in each cycle, one branch of infinite curve is followed in the two directions. In regions III and V, for t varying from $-\infty$ to 0 and then to $+\infty$, the point leaves B, goes to A and returns to B, whereas in region IV the point leaves B, goes to infinity and returns to B.

Note that, no matter what the value of Λ is, all models with a null minimum behave like the Einstein-de Sitter universe in the vicinity of that minimum, since their representative points are sufficiently close to point A. The cosmological constant may therefore be neglected when the universe is concentrated. The same holds for curvature effects at these same epochs, for the same reason as all the trajectories of types I, II, III, and V pass through A. These remarks will be useful in connection with the 3 K microwave background.

12.3 Age of the Universe

The age of the universe t_0 when Λ is not zero and when the radius goes through zero is given by solution (12.9), with y varying from 0 to 1:

$$H_0 t_0 = \int_0^1 dy/\sqrt{\Phi(y)} \tag{12.12}$$

which is a function of q and λ. It has been calculated by TOMITA and HAYASHI and is represented graphically in Fig. 12.3. For H_0 fixed, t_0 is all the smaller as λ and q become larger. t_0 is equal to $1/H_0$ all along one branch of a curve leading to the point $\lambda = q = 0$.

For the special case of the Eddington-Lemaitre models the universe has emerged from the state of the Einstein universe in the infinite past and its age is infinite. Below the curve $H_0 t_0 = \infty$, the models have a non-zero minimum.

Our universe is probably located in a band on the (q,λ) diagram between the curves $H_0 t_0 = 1$ and ∞, from the facts that the globular clusters formed about $12 \cdot 10^9$ yr ago and that the 3 K microwave background indicates that the universe has been very condensed. Therefore the cosmological constant ought to be positive, and space ought to expand indefinitely; but at this stage one is still unable to say anything about the sign of the curvature.

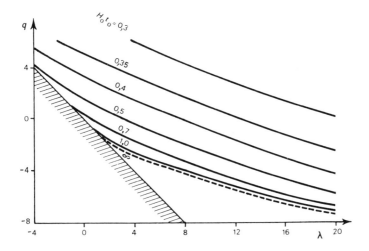

Fig. 12.3. Age of the universe

12.4 The Hubble Diagram

The Hubble diagram, the relation between the redshift and bolometric magnitude, has been calculated by TOMITA and HAYASHI for the case of positive cosmological constant and spherical space. These calculations were made by substituting in equations (11.6) giving z and (11.21) giving m_{bol}, the functions R(t) taken from (12.9).

Results are given in Fig. 12.4 for $\lambda = 40$ and various values of q. There is an astonishing deviation from the classical linear relation between log z and m_{bol}, in the sense that it returns back to the left. This is due to an effect of concentrating light rays when an object is near the observer's antipole: more rays reach us, and this concentration may compensate the loss of brightness due to the redshift. Taking expression (11.19) for the flux received f_o, we have $f_o \propto 1/\zeta^2 r_e^2$; near the antipole $r_e \to 0$, while ζ may remain at relatively modest values, from the fact that, in the Lemaitre models, R(t) may coast for a long period around a constant value (see the small central graph in Fig. 12.1). Thus f_o increases and m_{bol} decreases.

Computer calculations have been made for several other cases by GLANFIELD and REFSDAL, STABELL and DE LANGE.

In Fig. 12.4 the observations of HUMASON, MAYALL and SANDAGE for brightest galaxies in clusters are plotted. These observations show that $\lambda < 30$ or, from the definition (12.2) of λ,

$$\Lambda < 90 H_o^2. \tag{12.13}$$

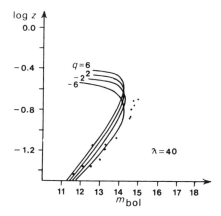

Fig. 12.4. The Hubble Diagram

With $H_0 = 80$ km sec^{-1} Mpc^{-1} we find

$$1/\sqrt{\Lambda} > 1.3 \cdot 10^9 \text{ yr} \quad \text{and} \quad c/\sqrt{\Lambda} > 0.4 \text{ Gpc},$$

this time and length indicating that the cosmological constant only has a significant effect on very large scales.

The Hubble diagram for cluster galaxies shows that our universe could be situated in the band of the diagram defined by $\lambda = 0$, $q = 0$ to 4 and $\lambda = 20$, $q = -4$ to 0.

In combining these results with those for the age of the universe, the observed value of the density, and the presence of the 3 K microwave background, it is probable that the representative point in the (q,λ) diagram is actually located in the neighbourhood of the segment joining $q = 0$, $\lambda = 0$ to $q = -1$, $\lambda = 1$. This region has been calculated in detail by HEIDMANN and NIETO. The universe would then probably be indefinitely expanding and rather hyperbolic. But even better observations must be made in order to carry out these estimates with any confidence.

13. Cosmological Horizons

In Chapter 11 we have seen that, in the Einstein-de Sitter universe, the number of observable objects is limited even though the space is infinite Euclidean. This happens because at a certain "distance" the spectral ratio becomes infinite and the photons then no longer bring us energy. The observations reach to a "horizon" due to cosmological expansion.

In the calculations the length $R_o r_e$ crops up. For the Einstein-de Sitter universe this length is equal to $2(c/H_o)(1 - \zeta^{-1/2})$ which, for $\zeta = \infty$, gives the limiting value $2(c/H_o) = 7.5$ Gpc, for $H_o = 80$ km s^{-1} Mpc^{-1}. This length limit gives an estimate of the "distance" at which we find the *cosmological horizon*, but it is necessary to know precisely the meaning of this term in advance of any questions that spring to mind. Among these questions, some have no definite meaning, especially those where the notion of distance to an object on the horizon, or even beyond the horizon, enters, those involving the concept of the velocity of the object in question, this speed being the derivative of a distance with respect to a time, but what distance and which time: at the moment of emission, detection..., time registered on the observer's own clock or a clock at the point of emission? Thus we shall remain in this chapter in close contact with the notions and parameters well-defined in the previous chapters and possessing certain observational interest.

The subject of cosmological horizons was introduced by BECQUEREL and COUDERC and developed by RINDLER. The easiest way to represent them is as a spherical surface in comoving coordinate space, which progresses in the course of cosmic time.

13.1 Particle Horizon

Equations (11.2) relating the time of emission t_e of a photon to the comoving radial coordinate r_e of the emitter, in the reduced sinusoidal representation, may be inverted and written

$$r_e = \mathscr{S}\left[c\int_{t_e}^{t_o} dt/R(t)\right] \qquad (13.1)$$

where the symbol $\mathscr{S}$ represents sin, 1, or sinh according to $k = 1$, 0 or -1.

By definition the *particle horizon* is the finite limit r_A, if such exists, of r_e as t_e tends to the lowest (inferior) possible limit t_{inf} in (13.1). We specify that, for the lowest possible limit of t_e, we take the epoch of the first zero minimum for R encountered in going back into the past; if there is no such minimum, the limit is $-\infty$.

The particle horizon is then the surface of a sphere defined by $r = r_A$ in the space defined by $t = t_0$ in spacetime. This is also the surface of a sphere $r = r_A$ in comoving coordinate space.

For example, for a universe with a zero minimum, $R(t) = R_0(t/t_0)^n$, with $n < 1$, $t_{inf} = 0$ and

$$r_A = \mathscr{S}[ct_0/(1 - n) R_0]. \tag{13.2}$$

The Einstein-de Sitter case corresponds to $n = \frac{2}{3}$ and $\mathscr{S} = 1$, so

$$r_A = 3ct_0/R_0.$$

For $n \geq 1$, r_A does not exist because the integral (13.1) diverges. For a universe with non-zero minimum, for example $R(t) = a + bt^2$, r_A is similarly calculated with $t_{inf} = -\infty$.

Properties of the Particle Horizon

1) Objects nearer than r_A are observable at time t_0, those further are unobservable. This follows by definition; in particular, if the universe had had a zero minimum, photons that could have left an object possibly existing before that minimum would have been thermalised and lost.

For spherical space we must watch for inconveniences arising from the reduced sinusoidal representation, the use of which is improper if one is considering a domain that spreads beyond the equatorial surface: for the same r, in any direction, there are several corresponding points. In this case it is better to use the central representation.

2) r_A increases in the course of cosmic time. In effect $R(t)$ is positive and hence the integral (13.1) increases if t_0 increases. Consequently one sees more and more objects.

3) Once an object is observable it stays that way. This occurs because r_A increases. Therefore, in the course of time, objects appear but none disappears.

4) An object is observed for the first time in the state in which it was at the epoch t_{inf}. It is observed then with an infinite spectral ratio if the universe has a zero minimum, and a zero spectral ratio if the universe has contracted from an infinitely dilute state in the infinite past and if, to be sure, r_A exists. In the first case, objects make their appearance with an extreme redshift and in the second case a blueshift. In this sense we may say that objects which cross the particle horizon to become visible, which they do at the speed of light, seem to recede in the first case and to approach in the second case.

We may illustrate the various properties of the particle horizon diagrammatically. To make the representation easier suppose the space has just two dimensions, take comoving coordinates r and θ and use the reduced sinusoidal representation. Take t along an axis at right angles to the (r,θ) plane; then equation (13.1) represents the tracks of photons which reach us at G_0 at time t_0, and these paths make a conical sheet, the light cone of G_0. This light cone is drawn in Fig. 13.1 for a universe with zero minimum - R(0) = 0. In this case the light cone becomes tangential to the

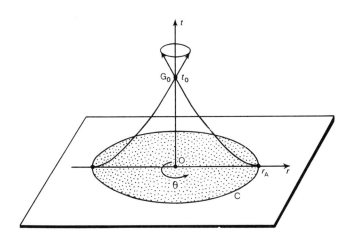

Fig. 13.1. Cosmological particle horizon

plane t = 0 for, in the case of photons, we have θ = constant and $ds^2 = 0$, therefore

$$c^2 dt^2 - R(0)^2 dr^2/(1 - kr^2) = 0$$

and thus dt/dr = 0. The particle horizon is the circle C: $r = r_A$, in the comoving space. All points of space on this circle C are being seen in fact for the first time; all points inside have already come into view. We may say therefore, in terms of *events*, in other words points of spacetime, that events situated on the anterior light cone are actually being seen, and events located inside this cone have been seen.

We take as an example the spherical Friedmann model. Equations (10.43) give

$$r_A = \sin \delta_0,$$

where δ_0 is the actual development angle. For simplicity we use the central representation of Section 9.2, with radial coordinate ω, represented by the model of Fig. 9.2. We then have $\omega_A = \delta_0$. But, following equation (10.48), $\delta_0 = \cos^{-1}(1 - q_0)/q_0$ which allows us to say: if $q_0 = \frac{1}{2}$, the particle horizon is at the point of observation; at this instant the expansion commences, $t_0 = 0$, as one may see from the q-marks in Fig. 10.2. For $q_0 = 1$, $\omega_A = \pi/2$, the horizon is on the equator; one can see half the spherical space; this happens after a time equal to about one-tenth the oscillation period of the universe. For $q_0 = \infty$, we have $\omega_A = \pi$: the horizon is at the antipole, the whole of space is visible, this happening when the universe is at the maximum expansion, after half a cycle time. Later ω_A gets larger than π and one starts to see space again, for the second time; after a cycle time one sees the observer "from behind".

We remark here that, if space is elliptical (see Section 8.8), all space is visible for $q_0 = 1$ and the observer is seen from behind for $q_0 = \infty$; after a complete cycle space is visible four times. These remarks about the differences in the appearance of spherical and elliptical space are not as academic as they seem at first sight. The differences are apparent as nearby as the equator; for the Einstein universe (Section 12.2), which gives order of magnitude, the radius of curvature is 19 Gpc and the equator is at 30 Gpc. Radio telescopes, if we interpret their results in the Einstein-de Sitter framework, have already penetrated to 3 Gpc, which is not that negligible compared to 30 Gpc; one can but hope that techniques or novel discoveries will let us fill the gap one day.

13.2 The Event Horizon

By definition we shall term the *event horizon* the limit, if it exists, r_L of r_A for $t_0 \to \infty$. The event horizon embodies all points of space with comoving coordinates which would be observable in the infinite future. If the universe passes through a zero minimum in the future, one may decide that, as far as we are concerned, the future does not stretch beyond the catastrophe in which we are wiped out, but mathematically the definition still makes sense. Therefore

$$r_L = \mathscr{S}\left[c \int_{t_{inf}}^{\infty} dt/R(t)\right]. \tag{13.3}$$

The diagrams in Fig. 13.2 represent the particle and event horizons in the same coordinates as Fig. 13.1 for $t_{inf} = 0$ or $-\infty$. We shall consider a worldline U of an object with fixed comoving coordinates, such that its radial coordinate r is less than the particle horizon r_A at time t_0. Such an object, visible at t_0, remains always visible to us, even though we would not observe its own history for times later than t'; there will be an infinite slowing down of the "movie" of this object such that its life seems dragged out near the asymptotic time t'.

The conical sheet - for the two-dimensional space employed in Fig. 13.1 - formed by light paths which would reach us at $t = \infty$ is called the *limiting cone*; its equation is

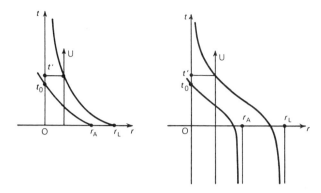

Fig. 13.2. Cosmological particle and event horizons

$$r = \mathscr{S}\left[c\int_t^\infty dt/R(t)\right]. \tag{13.4}$$

This cone discriminates observable events - points of spacetime inside the anterior cone - from those that are forever unobservable.

We note that it is possible to have a limiting cone without an event horizon. For example, if $R = R_0(t/t_0)^n$ with $n > 1$, the limiting cone is given by

$$r = \mathscr{S}\left[\frac{ct_0}{(n-1)R_0}\left(\frac{t_0}{t}\right)^{n-1}\right],$$

but r_L is infinite; in this case one sees the whole of space but one does not see its entire history.

It is also possible for there to be neither a limiting cone nor an event horizon, if the meridian of the limiting cone is shifted to the right at infinity, for example; this is the case of the Einstein-de Sitter model. One would then see the whole of space as well as all of its history.

13.3 The Absolute Horizon

Up to now we have assumed that the observer has fixed comoving coordinates, $r = 0$, and we have seen that space could not have been observed in this state beyond the event horizon when this exists. It is quite clear that, if we abandon this restriction of immobility and if we then move a given distance in a given direction, the limiting cone and the event horizon would be moved in the same direction. If, then, we want to extend our domain of exploration in the universe, we have to move.

In leaving G_0 at t_0, in the (r,t) diagram of Fig. 13.3 which corresponds to a model with zero minimum, one may go to any point inside the posterior light cone $\mathscr{P}$. In leaving as early as possible - at $t_0 = 0$ - and going as fast as possible - at the

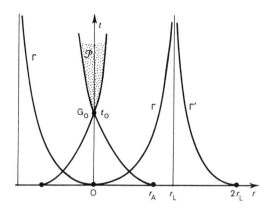

Fig. 13.3. Absolute cosmological horizon

speed of light - one can reach any point of the sheet Γ, the posterior light cone of the origin O, obtained by tracing the limiting anterior cones that correspond to points situated at r_L in the directions of motion of the observer.

If we decide to go along the right part of Γ, we would, from this region, in the infinite future, discover space as far as the coordinate $2r_L$ at most, as is indicated by the light cone Γ'. This limit $r_{ab} = 2r_L$ defines the *absolute horizon*. No point with fixed comoving coordinates can be observed beyond the absolute horizon. A point on the absolute horizon may be observed provided we leave in its direction at the origin of time, at the speed of light; and this, evidently, is disadvantageous to points located in other directions.

The absolute horizon has merely an academic interest, because of the severe observational conditions.

But it shows that, in the universe, parts separated from each other by r_{ab} or less can exchange messages or have interactions.

13.4 The Determination of Horizons

Horizons may be most readily determined via the central representation (Section 9.2). In this case equation (13.1) may be written

$$\omega = c \int_{t}^{t_0} dt/R(t), \qquad (13.5)$$

ω being the radial coordinate. For the zero pressure universes the equations of Chapter 12 give, with the same notations,

$$\omega = \frac{c/H_0}{R_0} \int_{y}^{1} dy/y\sqrt{\Phi(y)}. \qquad (13.6)$$

The relevant calculations have been made by STABELL and REFSDAL.

In particular, the event horizon ω_A is given by replacing t by t_{inf} in the limits of the definite integral (13.5), and therefore, for the zero minimum models of Chapter 12, y by 0 in the limits of equation (13.6). One can map out, in the (q,λ) diagram of Chapter 12, curves corresponding to ω_A equal to some value or other. Figure 13.4 gives the curves e and a whose points represent the universes - here spherical - for which the event horizon is on the equatorial surface and at the antipole, respectively. If our universe is in the neighbourhood of the point A, the equator is then still beyond the event horizon.

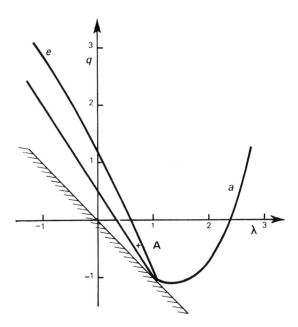

Fig. 13.4. Universes in which the event horizon is on the equator (e) or at the antipole (a)

For the Einstein-de Sitter universe, the direct calculation in (13.6) yields

$$r_{A.EdS} = \omega_{A.EdS} = 2\frac{c/H_0}{R_0}.$$

Here $R_0 = at^{2/3}$, a being a constant. Thus $H_0 = 2/3t_0$ and $r_{A.EdS} = 3ct_0^{1/3}/a$: the radial comoving coordinate of the horizon advances like $t_0^{1/3}$ in comoving space. But, in length, for the actual epoch t_0, the event horizon is $R_0 r_{A.EdS} = 2c/H_0 = 3ct_0$ and it advances linearly with t_0.

Part IV
The Metagalaxy in the Relativistic Zone

Having familiarised ourselves with spaces of constant curvature and relativistic models of the universe, we can tackle the study of observations in the relativistic zone of the metagalaxy. Of course, there is no precise boundary between the classical Metagalaxy and the relativistic Metagalaxy. In fact, in the first chapter of this part we ask again a question which we have already considered in the first part: the relation between the apparent bolometric magnitude of galaxies and their spectral shift, or the Hubble diagram.

14. The Hubble Diagram for Galaxies

The Hubble diagram which SANDAGE obtained for the galaxies (Fig. 3.1) reaches as far as a redshift of 0.5, already a relativistic value. The calculations in the third part of this book have shown that, from this value, relationships between apparent bolometric magnitude and redshift begin to differ, one from another, as the deceleration parameter q_0 varies, as Fig. 11.4 shows, for a cosmological constant $\Lambda = 0$. For $\Lambda \neq 0$, the differences may even be more perceptible which has made it possible to obtain an upper limit for Λ (Chapter 12).

The Hubble diagram obtained by SANDAGE is practically linear to the end, which, for $\Lambda = 0$, corresponds to $q_0 = 1$. In 1978 KRISTIAN, SANDAGE and WESTPHAL measured 30 new galaxies with shifts beween 0.2 and 0.75. It seemed that a slight curvature would appear in the diagram, leading to $q_0 = 1.6 \pm 0.4$, with $\Lambda = 0$ and not taking into account the evolution of the luminosity of the galaxies in the past. This would imply a spherical, oscillating space (see Fig. 12.1). Nevertheless, if one imposes $q_0 \simeq 0$, it is possible to account for the curvature of the Hubble diagram by a diminution of 0.5 magnitude in the luminosities of galaxies in the last 5 billion years, a value compatible with the works of TINSLEY. In this case the universe would be hyperbolic and in indefinite expansion.

We can see that the evolution of the universe and the evolution of galaxies are closely tied in the relativistic zone and that it is very difficult to untangle them.

15. Distant Intergalactic Material

The observation of intergalactic material in the relativistic zone proves to be very difficult and has only yielded upper limits, which we have already taken into account in Chapter 5 on the density of the universe. For convenience, we will divide this chapter into two sections, one devoted to neutral atomic hydrogen (and in addition, molecular), and the other to ionized hydrogen.

15.1 Neutral Hydrogen

Even if the 21-cm line of neutral atomic hydrogen is the easiest to observe, the strongest line is Lyman α, corresponding to the transition between the first excited state and the ground state. Its rest wavelength is 1216 Å, unobservable on the Earth, since our atmosphere absorbs radiation below about 3200 Å. The Lyman α line is not observable from Earth, unless the emitter has a spectral ratio ζ, large enough to lengthen the wavelength beyond 3200 Å, that is $\zeta > 2.6$.

In principle, if atomic hydrogen gas is distributed in space, either diluted in the universe or concentrated in a cluster, it should be possible, as in the case of the 21-cm line, to observe Ly α in emission. In fact, atomic hydrogen is too rare for it to be observed in Ly α emission on the ground.

On the other hand, if it lies between us and a very distant source of radiation, it might be seen in absorption. Quasars are possibly such sources, if one attributes to them the distances which cosmological models imply from their spectral ratios. The situation is quite similar to that studied in Chapter 4, but it must be treated with relativistic formulae.

The case of diluted hydrogen in space - In this case, one should observe a dip in the continuous spectrum of the distant source, between $\lambda_r = 1216$ Å and $\zeta_e \lambda_r$, ζ_e being the spectral ratio of the source. According to the calculations of BAHCALL, SALPETER, GUNN and PETERSON, the optical depth of the dip is, for Friedmann models, with $H_0 = 80$ km s^{-1} Mpc^{-1}

$$\tau = 4 \cdot 10^{10} n(\zeta_r)/\zeta_r \sqrt{1 - 2q_0 + 2q_0\zeta_r} \tag{15.1}$$

where $n(\zeta_r)$ is the concentration of hydrogen atoms per cm^3 at a distance ζ_r.

OKE's observations of quasars around $\zeta = 3$ have given negative results, with $\tau < 0.5$. Thus $n(3) < 3 \cdot 10^{-11}$ cm^{-3} which corresponds to a density of intergalactic atomic hydrogen, in the distant zone, less than $5 \cdot 10^{-35}$ g cm^{-3}. Ultraviolet observations of the nearby quasar 3C 273 from above the atmosphere lead to an even lower limit for the actual concentration:

$$n(1.2) < 1.5 \cdot 10^{-12} \text{ cm}^{-3}.$$

Quite analogous results are obtained for molecular hydrogen from its Lyman band situated around 1000 Å, leading to $n(3) < 4 \cdot 10^{-9}$ molecules per cm^3 or a density less than 10^{-32} g cm^{-3}.

The case of hydrogen concentrated in clusters - BAHCALL and SALPETER have also considered the case where the hydrogen is distributed within clusters of galaxies. If any such cluster lies in the path of photons reaching us from a quasar, each such cluster will produce an absorption line. The probable number of lines is given by equation (11.37). SALPETER calculated their optical depth; in the Einstein-de Sitter case, with $H_0 = 80$ km s^{-1} Mpc^{-1}, we have

$$\tau = 2 \cdot 10^{12} n\Theta/v_g \tag{15.2}$$

where n is the concentration of hydrogen atoms per cm^3 in the cluster, v_g their dispersion in km s^{-1} and Θ the linear diameter of the cluster in Mpc. Then for $\tau = 0.5$, $v_g = 1000$ km s^{-1} and $\Theta = 2$ Mpc, one obtains $n = 10^{-10}$ cm^{-3}.

We can see the extraordinary sensitivity of this method; $n = 10^{-10}$ cm^{-3} corresponds to a density of 10^{-34} g cm^{-3} and a mass of hydrogen in the cluster, assumed spherical, of only 10^7 $M_\odot$.

In the spectrum of the quasar PHL 957, with a spectral ratio of 3.69 measured from its emission lines, COLEMAN and his colleagues have observed more than 200 absorption lines from various atoms and corresponding to 20 different values of ζ, of which two, quite certain, are 2.80 and 3.31; the second may correspond to an intermediate galaxy. Other quasars also have several systems of absorption lines in their spectra. The widths of these lines are less than the resolution of the spectrograph, which is several tens of km s^{-1}; with such a small velocity dispersion it is unlikely that these lines are due to clusters; rather, they are due to galaxies; the great frequency with which these lines occur implies galactic haloes or discs at least 100 kpc in radius.

15.2 Ionized Hydrogen

So far, we have collected very little indication of the existence of intergalactic material. We still have to examine the case of ionized hydrogen, observations of which may be attempted optically, by radio waves, X-rays or γ-rays.

Optical observations - In a region of ionized hydrogen, by collisions with free, non-relativistic electrons, the photons are diffused by Thompson scattering with a cross-section $\sigma_T = 6.6 \cdot 10^{-25}$ cm^2. For ionized hydrogen distributed uniformly in space with an actual concentration n_0, the optical depth as far as spectral ratio ζ is, in the Einstein-de Sitter case,

$$\tau = \frac{2c\sigma_T n_0}{3H_0}(\zeta^{3/2} - 1). \tag{15.3}$$

For $n_0 = 10^{-5}$ cm^{-3}, corresponding to the density ρ_{EdS}, the equation (15.3) gives $\tau \simeq 0.2$ for $\zeta = 3$. So, Thompson scattering due to intergalactic ionized hydrogen may be detectable in quasars for the indicated concentration. This observation, which is still difficult, has so far not been possible.

Radioelectric observations - Radio waves are absorbed by ionized hydrogen. The formulae are those for thermal radiation in a gas with some electron temperature T_e; the optical depth, governed by ion-electron collisions, is given, out to spectral ratio ζ, by

$$\tau \nu^2 T_e^{3/2} = 0.035 \, n_0^2 \frac{c}{H_0}(\zeta^{9/2} - 1), \tag{15.4}$$

in the Einstein-de Sitter case, ν being the frequency.

For large ν, τ is small. The principle of the observations is to register the spectra of fairly distant radio sources, and to see whether, at low frequencies, there is a decrease of emission from the more distant ones, in comparison with the nearer ones. ERICKSON and CRONYN have thus observed eight radio sources out to $\zeta = 1.4$, leading to a negative result: $\tau < 0.2$, that is

$$n_0 < 4 \cdot 10^{-7} T_e^{3/4} \text{ cm}^{-3}, \tag{15.5}$$

not a very interesting limit since, if the hydrogen is ionized, T_e is at least 10^4 degrees.

Observations in X- and γ-rays - Some extragalactic sources of X- and γ-rays have been discovered, such as *Virgo X-1*, associated with the radio galaxy *Virgo A*, a giant elliptical in the Virgo cluster; others are associated with Seyfert galaxies (NGC 4151) or quasars (3C 273).

As well as these discrete sources, a general background has been observed whose spectrum decreases rapidly from 250 eV to 4 GeV, the energy limit of X- and γ-ray photons which are actually observable. In the neighbourhood of 10 keV, SCHWARTZ has shown that the radiation is isotropic to better than 4 per cent, indicating a general extragalactic origin.

The origin of this diffuse X- and γ-radiation must be complex because of the changes in gradient observed in the spectrum. Equation (15.3) shows that it may come from regions very deep in space, where large clusters of galaxies and Seyfert galaxies, which may radiate up to 10^{45} erg s^{-1} between 2 and 10 keV, and quasars up to

10^{46} erg s^{-1} between 50 and 500 MeV, must contribute an important part of this diffuse background.

The rest may come from the interaction of relativistic electrons with low energy photons of the 3 K microwave background which, by the inverse Compton effect, give them a high energy, transforming them to X-ray or γ-ray photons. However, the origin of these electrons is not clear.

Extragalactic X- and γ-rays certainly have considerable cosmological importance; but in the face of the broad range of possible theoretical interpretations, particularly the accretion of matter by massive black holes, it is necessary to obtain more detailed observations.

16. Radio Galaxies and Quasars

16.1 Basic Data

In this chapter we tackle an area of knowledge about the universe essentially covered by radio astronomy but where, nevertheless, optical astronomy has played and will continue to play a considerable part.

We will be dealing chiefly with radio sources, characterized first by their positions in the sky, and then by their flux densities s at wavelength λ, at which they are observed. These flux densities will be expressed in the unit of flux (density), the *Jansky* (Jy), equal to 10^{-26} Wm^{-2} Hz^{-1} and the wavelength in centimetres. The study is here limited to sources with galactic latitudes outside the range ± 10 or $20°$ which, in practice, ensures that they are all extragalactic.

The great breakthrough in extragalactic radio astronomy was made between 1951 and 1954 by G. SMITH, BAADE and MINKOWSKI when they identified the strongest radio source in the sky, *Cygnus A* (s_{168} = 8100 Jy) with a very peculiar galaxy having a radial velocity among the largest known at the time: 16800 km s^{-1}, being equivalent to a spectral ratio ζ = 1.05. In 1961, the seventh radio source in order of intensity (s_{168} = 73 Jy), *Bootes A*, was identified with a galaxy almost at the limit of visibility of optical telescopes with ζ = 1.46, a record for a galaxy, held for ten years. As radio telescopes are actually capable of detecting radio sources of 0.001 Jy, we grasp the enormous scope which radio astronomy must have in the exploration of space, thanks to the existence of radio galaxies.

In 1963, M. SCHMIDT correctly deciphered the optical spectrum of a 12th magnitude "star" found at the position of a radio source, 3C 273, and found a spectral ratio corresponding to a radial velocity of 50000 km s^{-1}. A few years later, other "quasi stellar sources" were found, with enormous spectral ratios: 3 for 3C 9; the record is held (1979) by the 18th magnitude quasar OQ 172 with ζ = 4.53.

The interpretation of such large redshifts for objects so bizarre and new as the quasars is still a subject of controversy. Many astronomers think that these redshifts must be interpreted in terms of cosmological models of the universe, that is that they imply "cosmological" distances for these objects, greatly exceeding those of the most distant galaxies observed. STOCKTON has shown with an unbiased sample of 27 quasars that 8 of them are associated with at least one galaxy having the same radial velocity to within 1000 km s^{-1}; the probability that this would happen by chance is less than 10^{-6} and essentially proves the cosmological nature of the spectral shifts of these quasars.

On the other hand, some indications lead one to believe that the spectral shifts may have another origin. ARP, especially, found several troubling cases which apparently seem to require some fundamental changes in physics. In particular, according to a theory by HOYLE and NARLIKAR, the masses of elementary particles may not be invariable.

16.2 Number Counts

The principal catalogues of radio sources have been made by RYLE and his collaborators at Cambridge, at metre wavelengths: the 3C to $s_{168} = 9$ Jy, the 4C to $s_{168} = 2$ Jy and the 5C down to $s_{74} = 0.01$ Jy. In centimetre waves the catalogues go down to 0.001 Jy.

Simple knowledge of the flux densities s in a catalogue allows a curve N(s) to be drawn, giving the number of radio sources more intense than s. The greatest amount of work in this area has been done by RYLE and his team, at first at a wavelength of

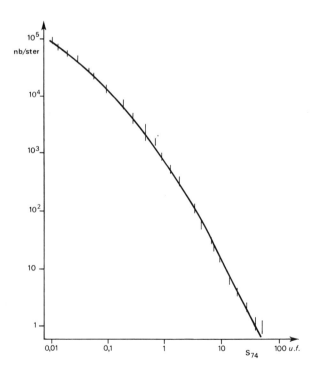

Fig. 16.1. Number counts of radio sources

168 cm, over a range of flux density of 100 : 1, and later at 74 cm over a range of 5000 : 1. Figure 16.1 gives the curve obtained by POOLEY and RYLE at 74 cm, with the error bars. For $s_{74} > 4$ Jy, the slope of the curve on a logarithmic scale is -1.85, then increases progressively to -0.8 for $s_{74} = 0.01$ Jy.

This is a major result from radio astronomy, of great importance for cosmology. For a static Euclidean universe, uniformly populated with invariable radio sources, the slope would be -1.5, corresponding to the slope 0.6 in the galaxy counts as a function of magnitude, equation (2.2). On the other hand, in all the relativistic models of the universe we have studied in Chapter 11, the curve must have an algebraically greater slope, by a mixed effect of the expansion and the geometry. In Fig. 16.1 there is an excess of weak radio sources; one possible interpretation is the following: on average, the weak radio sources are further away and so are seen as they were in the distant past. In the past, radio sources would have been more numerous than now. OORT was one of the first people to point out that this implies that important evolutionary effects enter the relation between the population of radio sources and cosmic time.

16.3 Distribution and Luminosity Function

If, in a catalogue made at a frequency ν_0, which is complete to a limit of flux density s_1, one knows the spectral ratios ζ of all the radio sources, one may calculate for each, by means of the formulae in Chapter 11 (equations (11.27), (11.28) and (11.30)) the power $F(\nu_0 \zeta)$ emitted per steradian per hertz at the frequency $\nu_0 \zeta$, a quantity called, for short, *radio luminosity*. For spectral ratios near 1, the radio luminosity is practically that at ν_0; for ζ greater than 1, if one knows the form of the spectrum - its spectral index in the case of a power law spectrum - one may evaluate its radio luminosity at ν_0.

Under these conditions, from this catalogue, one may then construct the *distribution of radio luminosity* $n(F, s_1, \nu_0)$, giving the number of radio sources per steradian and per interval of radio luminosity. If $\rho(F, \nu, t)$ is the *radio luminosity function*, giving the number of radio sources at cosmic time t, per unit volume and per interval of radio luminosity at frequency ν, the relation between the distribution and function of radio luminosity is

$$n(F, s_1, \nu_0) = \int_0^{r_1} \rho(F, \nu_0 \zeta, t) R(t)^3 r^2 dr / \sqrt{1 - kr^2} \ . \tag{16.1}$$

This relation is found by summing the radio sources up to the comoving radial coordinate r_1 - in reduced sinusoidal form (9.18) - such that, at r_1, the radio luminosity F appears with the flux density limit s_1; r_1 is then given by equation (11.27).

If radio sources are sufficiently near, the expansion of the universe, the curvature of space, and the evolution of radio sources are negligible and (16.1) gives simply

$$n(F,s_1,\nu_0) = \rho(F,\nu_0,t_0)(F/s_1)^{3/2}/3. \tag{16.2}$$

Sometimes *apparent* and *absolute radio magnitudes* are used, defined by

$$\left. \begin{array}{l} m = -53.45 - 2.5 \log_{10} s \\ M = 34.00 - 2.5 \log_{10} F \end{array} \right\}, \tag{16.3}$$

where s and F are in W m^{-2} Hz^{-1} and W Hz^{-1} ster^{-1}. Then (16.2) becomes

$$\rho(M,\nu_0,t_0) = 3 \cdot 10^{24} \cdot 10^{0.6(M-m_1)} n(M,m_1,\nu_0) \tag{16.4}$$

per Gpc^3 and magnitude.

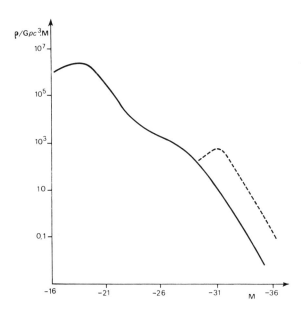

Fig. 16.2. Radio luminosity function

From the distributions of radio luminosity, which depend greatly on the catalogues, the radio luminosity functions may be deduced using (16.2), which are physically much more interesting. The solid curve in Fig. 16.2 gives the number of radio sources, per Gpc^3 and per magnitude at a wavelength of 168 cm, according to the work of CASWELL and WILLS, for H_0 = 80 km s^{-1} Mpc^{-1}. One sees that strong radio sources are the least numerous in space.

In the case of large spectral ratios (16.1) must be treated exactly. To do this we express r and t as functions of ζ using a model of the universe. For the Einstein-de Sitter model we obtain

$$n(F,s_1,\nu_0) = \frac{8}{3}(c/H_0)^3 \rho(F,\nu_0,t_0)(1 - \zeta_1^{-1/2})^3 \tag{16.5}$$

where it remains to express ζ_1 in terms of F and s_1 by (11.27); for this a good knowledge of the spectrum $F(\nu)$ is needed. In (16.5), when writing t_0 in ρ, we have assumed that there was no evolution of radio sources during cosmic time t.

In fact, ρ may evolve with t, as we have indicated à propos the number counts, and the calculations are then much more complicated. HEIDMANN and LONGAIR have shown that it is possible to construct models where radio sources evolve which take account of the number counts. Although these models only represent a first approximation, they indicate the principal characteristics of the type of evolution which must take place. The steep slope of the curve (log N, log s) for the radio sources may be explained by a very rapid decrease in the density of objects as a function of cosmic time, or by a decrease in their radio magnitude. In reality, there is probably a mixture of evolution in radio luminosity and evolution in density.

In identifying photographically almost all of the 60 most intense radio sources in the 3C catalogue, LONGAIR split them into two groups, 23 quasars and 33 or 35 radio galaxies, and has shown that the slope of the number count is -2.2 for quasars, -1.9 for powerful radio galaxies, and -1.1 for weak ones. He deduced from this that the density (in covolume) of quasars and of powerful radio galaxies evolves with cosmic time according to $\rho \propto \exp(-10\ H_0 t)$, for $\Lambda = 0$, $q_0 = 0$. This corresponds to an enormous variation and a lifetime for these populations of the order of only one billion of years.

The dotted curve in Fig. 16.2 gives an example of the radio luminosity function calculated for $\zeta = 3$.

SCHMIDT has shown by a method of maximum accessible volume that among the quasars those with flat radio spectra $F(\nu)$ (spectral index > -0.2) evolve less quickly than those with steep spectra (index < -0.2).

In the optical field, BRACCESI has compiled a catalogue of 300 optical quasars and found for their number N, over the whole sky, down to magnitude B

$$\log_{10} N = -1.80(-B/2.5) - 8.7 \tag{16.6}$$

for B in the range 17.5 to 19.4; as -B/2.5 plays the same part as $\log_{10} s$, the slope of the number count of optical quasars is the same, -1.8, as that obtained by RYLE for bright radio sources. BRACCESI and SCHMIDT accounted for this slope by similar evolution.

The study of the apparent diameters of radio sources also leads to evolution. From the actual distribution in diameter, obtained in the comparatively nearby 3C survey,

LONGAIR and POOLEY calculated the observed distribution in the 5C survey, which was deeper and thus further in the past, leading to an evolution in coconcentration of radio sources. This variation is less rapid than that obtained for strong radio sources; but here, in fact, we deal with radio sources of average radio luminosity, since the radio sources with large linear diameters - the most sensitive to this test - are more advanced in their own particular evolution and are less powerful.

Other tests seem to indicate that the linear dimensions of radio sources will have varied in the course of cosmic time, as ζ^{-1}.

It is very difficult, while we do not have adequate understanding of the physics of radio sources, to use their dimensions and power to survey the universe, as these are very dispersed.

These various facts appear to suggest that radio sources underwent important evolution during the course of cosmic time, in number and in radio luminosity, and that, from this point of view, the universe was very different in the past from its present state. That, added to the cosmological interpretation of the microwave background (see Chapter 17), goes against the steady state theory of BONDI, GOLD and HOYLE.

16.4 Isotropy of Extragalactic Radio Sources

The catalogues have been analysed by WEBSTER for the distribution of radio sources over the celestial sphere. There is no indication whatsoever that radio sources are not distributed independently and randomly on the sky. A single abnormal case has been pointed out by WILLIS, where an excess of a dozen strong radio galaxies are contained in a zone $4°$ in diameter.

The majority of radio sources which have been analysed must have ζ between 2 and 4, which corresponds to looking back seven-eighths of the way through the past history of the universe. The fluctuations are at most 3% on the scale of a cubic gigaparsec, and, furthermore, that limit comes from the limited number of radio galaxies contained in the catalogues, some 10^4. This isotropy is evidently primary data for cosmology along with the one of the microwave background.

It is likely that surveys to fainter levels will be less interesting, as the population of radio sources detected will be more and more dominated by nearer, less powerful objects, as we have remarked with reference to the distribution of radio luminosity.

16.5 Test of Closure

A new kind of test has been proposed by SOLHEIM: if space is spherical, and if the cosmological horizon is beyond the antipole, one would be able to observe the same radio source in two opposite directions, on condition that it radiates for a sufficiently long time, longer than the difference between the two light travel times.

PETROSIAN and SALPETER have worked on the problem of opposing images in various models of the universe, giving the relationships between the spectral ratios, the magnitudes, and the apparent diameters of the two appearances; local irregularities in the curvature of space due to galaxies and clusters of galaxies in the paths of photons may deviate the radiation and change the "focal length", thus breaking the two images into several neighbouring fragments.

SOLHEIM proposes a list of radio sources which are possible candidates, but the statistics are still insufficient to confirm that we really are dealing with radio sources whose images are reaching us from two opposing directions.

17. The Cosmic Microwave Background

If the interpretation which the majority of astronomers give to the 3 K cosmic background radiation is actually right, this radiation constitutes the second phenomenon of primary importance to cosmology, after the expansion of the universe. In effect, it proves that our universe has known an extremely condensed state in the past, emerging abruptly from a quasi-point-like state by an inverse process to gravitational collapse.

The uniform models of the universe of General Relativity which we have presented in Chapter 12 all imply at some stage the value $R = 0$ of their radius of curvature or scale parameter, except those situated in the lower right of Fig. 12.1.

The possible existence of a cosmic background was foreseen in the forties. If the universe has been very condensed in the past, it must have been filled with radiation in thermal equilibrium at a very high temperature, the residue of which is now observed, very much cooled by expansion, in the shape of electromagnetic radiation with a low-temperature thermal spectrum.

The characteristics of the cosmic background allow the physical state of the universe to be calculated - density, pressure, temperature - back into the distant past, and the events of the first instants to be retraced, when it was thrown into a violent stage of nuclear reactions which, in a quarter of an hour, transformed 25 per cent of all the hydrogen into helium.

Furthermore, the cosmic background allows us to measure the velocity of the Sun, by means of the Doppler effect, with respect to a quasi-absolute system of reference. When the observations have become precise enough, this reference will be a great help in elucidating the dynamical problems posed by the local Metagalaxy, in particular the movement of the Local Group of galaxies relative to the *Virgo* cluster.

17.1 Description of the Cosmic Background

The cosmic background radiation was discovered in 1965 by PENZIAS and WILSON and by DICKE, PEEBLES and their collaborators, at wavelengths of several centimetres. Above 30 cm wavelength, the synchrotron radiation of the halo of our Galaxy becomes stronger than the cosmic background and begins to hinder the measurements, while under several millimetres, measurements from the ground are made very difficult by the radiation from the Earth's atmosphere. Nevertheless, measurements between 0.5 and 3 mm have been made from balloons. In the intermediate range, from 3 mm to 1 m, measurements of brightness by radio techniques are in agreement with a thermal spectrum at 2.7 ± 0.1 degrees Kelvin, that is, with *black body* radiation at T = 2.7 K whose brightness β is given by the Planck formula

$$\beta = 2hc^2/\lambda^5 (e^{hc/\lambda kT} - 1). \tag{17.1}$$

On logarithmic scales this spectrum presents a steep climb in submillimetre waves, a maximum around 1 mm, then a linear decline to large wavelengths. We note that there is agreement in absolute value; everything is as if we were enclosed in a perfect oven at 2.7 K. It is also worth noting that the measurements cover a considerable range of wavelengths of 2000 : 1 and that the maximum of the Planck spectrum in this range is apparent in the measurements, indicating that we are not dealing with *grey* radiation - that is, partial - at a higher temperature.

At shorter wavelengths measurements have been obtained by an indirect method suggested by FIELD, using interstellar molecules as thermometers. In the spectra of several stars one sees around 3874 Å two absorption lines due to an intervening cloud of interstellar gas; one is produced by the CN molecule in its fundamental state, the other by the same molecule in its first rotational state. To obtain this excited state photons of 2.6 mm wavelength are needed. The measurement of the relative intensities of the two lines gives the relative populations of the two states, and thus the density of photons present, which one may calculate from the temperature of the black body which produced it.

The measurements by THADDEUS and his collaborators on a dozen different stars have given temperatures between 2.7 and 3.0 K. Such a narrow distribution shows that the radiation responsible, the same at various points in our Galaxy, is independent of local circumstances and is thus the cosmic background observed from the Earth.

THADDEUS has attempted to observe other lines of CN, CH and CH^+ at 1.32, 0.56 and 0.36 mm, but has only obtained upper limits compatible with radiation at 2.7 K and clearly indicating a bend in the spectrum.

After several systematic series of observations of the cosmic background in centimetre waves, it has been shown that the radiation is isotropic, with great precision: one small anisotropy of 0.1 per cent over 360° of sky - to which we will return in the final paragraph - and nothing more than several thousandths on scales of 15° and 2'.

This remarkable isotropy precludes the possibility that the cosmic background has its origin in the Solar system, or in our Galaxy, or in the nearby Metagalaxy, because of the anisotropy of these systems. It also makes an origin in the cumulative emission of individual sources distribution in space improbable. Thus it lends weight to the interpretation of the cosmic background as residual radiation from the condensed phase of the universe.

The isotropy of the cosmic background implies that the expansion of the universe occurred in a very precisely isotropic way, from the earliest times, as THORNE has shown. If the expansion had been, for example, "cigar-shaped", that is more rapid in one particular direction, the photons would have been more spread out in this direction, and thus colder, and the cosmic background there would have a lower temperature.

Another interesting point about this almost perfect isotropy is that it means the cosmic background may be used as a quasi-universal system of reference. In effect, given the present low density of the universe, it may be estimated that these photons have travelled without interacting with matter for a very long time on the cosmic scale; they thus constitute a reference frame tied to the distant matter which last diffused them, the spectral ratio of which is estimated to be 8 at least. If the Sun travels with a velocity v relative to the radiation, the frequency of the observed photons in the direction of movement will be increased by the Doppler effect, and the temperature of the background will be raised in this direction by an amount ΔT such that $\Delta T/2.7 = v/c$.

The result of SMOOT and his colleagues from a U-2 plane, observing at a wavelength of 0.9 cm, lead to a slight anisotropy of 0.13 ± 0.02%, indicating that the velocity of the Sun, with respect to the cosmic background radiation, and thus with respect to the universe on a very large scale, is 390 ± 60 km s^{-1} towards 11.0 ± 0.6 hours of right ascension and 6 ± 10 degrees of declination.

This implies that our Galaxy travels relative to the reference frame defined by the microwave radiation - assumed to be really isotropic - at 600 km s^{-1} towards $l = 261°$, $b = 33°$. This result differs from the motion relative to the Metagalaxy given in Section 3.1 and poses a problem. Perhaps the microwave radiation is not in fact perfectly isotropic.

17.2 Cosmological Interpretation

If the cosmic background may be interpreted as residue from the condensed epoch of the universe, Section 10.6 gives us the equations we need to study it. In particular, according to equation (10.59), its temperature varies as $1/R(t)$ and its density, according to (10.55), as $1/R(t)^4$. Its present density, given by (10.58), has the value $5 \cdot 10^{-34}$ g cm^{-3} for $T = 2.7$ K.

Thanks to this data on the radiation and similar data relating to matter - density varying as $1/R(t)^3$ and present value $3 \cdot 10^{-31}$ g cm^{-3} according to (10.39) and (5.8) - one may reconstruct the variation of the physical properties of the universe in the course of cosmic time, in a remarkable shortened account of its history, and construct the evolutionary graph of Fig. 17.1.

As abscissae, below, are the values of $R(t)/R(t_0)$ on a logarithmic scale. The present corresponds to $\log (R/R_0) = 0$ and the past is at the left. The line labelled T represents the variation in temperature of the radiation on the logarithmic scale at the left, while the line ρ_r gives its variation in density with the logarithmic scale to the right, in g cm^{-3}. Finally, the line ρ_m represents the variation in the density of matter.

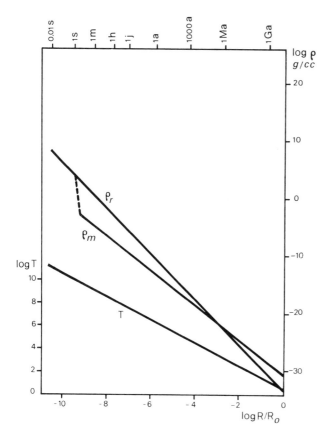

Fig. 17.1. Almost the complete history of the universe

We note that, when the universe was a thousand times smaller than it is now, the lines ρ_m and ρ_r crossed: after this point, matter predominates, before it, radiation. When there is a preponderance of matter, the equation of state p = 0 is valid, and one may use corresponding models of the universe; we remarked at the end of Section 12.2 that, while the universe is small, it is possible to neglect the cosmological constant and the effects of curvature, all models thus being close to the Einstein-de Sitter model. Formula (10.41) of this latter model may be used for R(t), where $R \propto t^{2/3}$. This allows the abscissa of Fig. 17.1 to be calibrated in cosmic time for the period when matter dominates; this has been done in the scale at the top, adopting at the right, the round figure of 10^{10} years for the age of the universe.

Similarly, while radiation was dominant, R(t) is given by equation (10.57):
$R \propto t^{1/2}$, by which the left of the scale may be calibrated, and gives

$$T = 1.5 \cdot 10^{10}/t^{1/2} \text{ K s}^{-1/2} \tag{17.2}$$

during the radiative period.

A hundredth of a second after the start of the expansion the temperature is 10^{11} degrees; the radiation density, 10^{12} g cm^{-3}, is such that it consists of an opaque gas of photons, electrons, and neutrinos in thermodynamic equilibrium. At such a temperature even atomic nuclei are unable to exist; matter proper only exists in the form of neutrons, and above all, as the more stable protons, with a density of the order of 10 g cm^{-3}, a value very much less than the densities existing in the interiors of white dwarf stars, for example; the nucleons are then in what is essentially a classical situation, particularly at this temperature, their speeds are not even relativistic.

A second later the temperature of the mixture has lowered and the average energy of the photons goes below a million electron volts, which allows electron-positron pairs to begin to annihilate and be transformed into radiation. This destruction of matter originally in the form of electrons is represented in Fig. 17.1 by the short dotted line.

The formation of helium - At ten seconds, the temperature falls below 10^{10} degrees and, by nuclear reactions, protons and neutrons may start to create deuterium. WAGONER, FOWLER and HOYLE have numerically calculated with the aid of cross-sections from nuclear physics the various nuclear reactions which then take place:

$$n + {}^{1}H \rightarrow {}^{2}H + \gamma$$

$$^{2}H + {}^{2}H \rightarrow {}^{3}He + n$$

and so on, up to the nucleus of magnesium.

In one minute the production of ^{4}He climbs drastically; that of deuterium reaches a maximum after several minutes, then decreases to a proportion by mass of 10^{-5} in a quarter of an hour, while ^{3}He rises to around 10^{-5}. ^{7}Li reaches a proportion of 10^{-8}, other nuclei being much less abundant, and heavy nuclei being practically absent.

At the end of a quarter of an hour the temperature is too low and the universe too diluted for reactions to continue. The essential result of these active minutes in the history of the universe is to transform the matter into 75 per cent hydrogen and 25 per cent helium.

Then the universe experiences a long quiet period. After 10,000 years the temperature gets below 10,000 degrees, and the gas which has up till now been ionized begins to pass into the neutral state and to decouple itself from the radiation. Afterwards the temperature keeps going down, the universe becomes darker and darker, bright red, dark red, black.

After 300,000 years the radiation loses its domination over matter, and millions of years later condensations appear in the gas, leading to the first galaxies, illuminated by the first generation of stars, after about a hundred million years.

Formation of the galaxies - EGGEN, LYNDEN-BELL and SANDAGE, whose theory rests particularly on the elongated shapes of the orbits of the stars in the galactic halo, think that the first stars formed during the gravitational contraction of the cloud from which our Galaxy formed, a rapid contraction taking a little less time than the present period of revolution of our Galaxy, some 100 or 200 million years.

PARTRIDGE and PEEBLES have outlined a model for the formation of galaxies. If one considers a homogeneous spherical cloud of radius r and mass $\mathcal{M}$, the Newtonian approximation may be applied to study its evolution; the variation of its radius is given by Newton's law: $\ddot{r} = -G\mathcal{M}/r^2$, which gives a cycloidal variation of $r(t)$; the cloud stops its expansion at the end of a time:

$$t_{max} = \pi\sqrt{r_{max}^3/8G\mathcal{M}}. \tag{17.3}$$

Taking for r_{max} the radius of the halo of our Galaxy, which may be considered as a relic of the maximum dimension of the protogalactic cloud, one obtains $t_{max} \simeq 1.5 \cdot 10^8$ years. After this time the cloud contracts and a flattened system results after about $3 \cdot 10^8$ years.

PARTRIDGE and PEEBLES think that a period of intense, bright star formation took place at the start of the contraction. This phase of ignition ought to make the galaxies observable in the near infrared with spectral ratios of the order of 30 and diameters of 30 arc sec. As the results of observation have been negative, SUNYAEV, TINSLEY and MEIER proposed a more intense burst of star formation at the end of the contraction. They extrapolated the properties of HII regions of the Orion type and found that these galaxies will have very characteristic spectra and that with spectral ratios from 3 to 30 they may have the appearance of slightly "woolly" quasars. The giant *clumpy irregular galaxies* discovered by CASINI and HEIDMANN, in which the clumps are a hundred times stronger than giant HII regions, may provide still better models for these primordial galaxies.

The hadron era - What happened during the first hundredth of a second? We have seen that, at a time

$$t_e \simeq 10 \text{ s},$$

the mean energy kT of the thermal radiation is sufficient to create electron-positron pairs $e^\pm$ with their corresponding neutrinos ν. If we go back in time, kT reaches the value $m_\mu c^2$, where m_μ is the mass of the muon, 207 times that of the electron, for

$$t_\mu \simeq 2 \cdot 10^{-4} \text{ s}$$

according to equation (17.2). Thus, for $t \lesssim 2 \cdot 10^{-4}$ s, we are in a period when, along with the thermal radiation, the leptons $\mu^{\pm}$, $e^{\pm}$, ν, responsible for weak interactions, exist.

In the same way, at a time

$$t_{\pi} \simeq 10^{-4} \text{ s}$$

kT becomes equal to $m_{\pi}c^2$, where m_{π} is the mass of the pion, the lightest of the hadrons, responsible for strong interactions. And so on; the further one goes back in time, the more the thermal radiation will be contaminated by particles of greater and greater mass, created in pair production: kaons, protons and antiprotons, neutrons and antineutrons, hyperons and antihyperons...

If one wishes to go back even earlier in time, one begins to run into difficulties because our understanding of particle physics becomes less and less certain progressively as their masses increase, because to study them they must be created, which requires more and more powerful accelerators. According to HAGEDORN the mass spectrum of created hadrons increases exponentially with mass, and the temperature of the radiation then tends towards a limiting value, a "hadron boiling point", of the order of 10^{12} K. The universe will then have this fixed temperature before

$$t_b \simeq 2 \cdot 10^{-4} \text{ s}.$$

Nevertheless, for small cosmic times the cosmological horizon is at a distance of the order of ct, as we have seen in Chapter 13. If the Compton wavelength $\lambda = \hbar/mc$ of a particle of mass m - a length which gives an order of magnitude of its dimensions - is greater than ct, that particle is not able to exist. Thus the particles of mass $m < \hbar/tc^2$ must be eliminated according to HARRISON, and the hadron states begin to depopulate for

$$t_h \simeq \hbar/m_{\pi}c^2 \simeq 10^{-23} \text{ s}.$$

Under this value the temperature of the universe will again rise as given by (17.2).

However, it should be said that our knowledge of physics is insufficient at the moment to tackle this period. Anyway, progress in the physics of quarks indicates that, when the density of the universe exceeds the nuclear density $2 \cdot 10^{15}$ g cm^{-3}, quarks should enter the scenario.

The quantum era - For t smaller still it appears that quantum effects must come into play. From the union of General Relativity, which involves the constants c and G, and quantum mechanics, where $\hbar$ comes in, one may construct a single universal length, the length

$$L_p = \sqrt{\hbar G/c^3} \simeq 10^{-33} \text{ cm} \tag{17.4}$$

discovered by PLANCK in 1899. The cosmological horizon is equal to the PLANCK length for

$$t_p \simeq 10^{-43} \text{ s} .$$

According to WHEELER, before this time, the whole universe may be governed by uncertainty relations that blur the geometry. He represents the allowed geometrical states of three-dimensional space E by points $\underline{M}$ in a superspace $\mathscr{S}$. Each point of $\mathscr{S}$ corresponds to a slice t = constant of spacetime, as given by solutions of Einstein's equations (Fig. 17.2). The evolution of the universe will thus be represented, in the non-quantum approximation, by a curve Γ of $\mathscr{S}$, for example by a succession of spherical spaces with curvature R(t) varying cycloidally, in the case of a positively-curved Friedmann universe. But, in the quantum treatment, this curve Γ would, in reality, be the path of a probability wavepacket, the waves being the solution of a wave equation in superspace, in the same way that the path of a particle in ordinary space is described by the evolution of a wavepacket derived from quantum mechanics. The path Γ would in reality have a fluffiness with the result that the geometry of space has, at every point, an uncertainty of order of the Planck length. In practice, these uncertainties are of no importance; but when the universe is very condensed, relatively important quantum effects could appear. In particular, for $t < t_p$, the universe could exist in several different quantum states, the fluffiness of the curve Γ having greater and greater importance, relatively, on the geometry of the universe, as the origin $\underline{0}$ (t = 0) is approached. Therefore we could not determine the state of

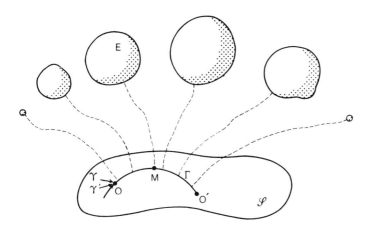

Fig. 17.2. Superspace

the universe for $t \to 0$. For the same reason we could not determine which path $\gamma, \gamma', \ldots$ it has followed before $t = 0$. It could even be that the universal constants $c, G, h, e, m_e, \ldots$, would be subject to changes during the passage through the very dense phase. If our universe is in an oscillating phase, it could, after the next collapse, be reborn with entirely different properties.

These considerations of quantum cosmology are most interesting and they will perhaps make an important contribution to the study of the universe. But, for the present, observational facts are practically non-existent for the era of cosmic time $t < 10^{-4}$ s.

In summary, the history of the universe may be broken down into successive eras. In order to give fair weights to their successive durations we express them, arbitrarily, as logarithms of the time $\tau = \log_{10} t$, t being in seconds.

1) The quantum era for $\tau < -43$.
2) The hadron era, uncertain for $-43 < \tau < -23$; at 10^{12} K (?) for $-23 < \tau < -4$.
3) The lepton era for $-4 < \tau < 1$.
4) The radiation era for $1 < \tau < 13$.
5) The gaseous era for $13 < \tau < 15.8$.
6) The stellar era for $15.8 < \tau < 17.5$, 17.5 corresponding to the present epoch. To put these ideas in another context, the "human era" corresponds to $17.4999 < \tau < 17.5000$. Our era seems rather small compared to the hadron era!

The origin of matter - In fact, many very important events took place in the first ten-thousandth of a second. OMNÈS has supposed that the matter that actually exists in the universe is the residue of matter and antimatter which contaminated the thermal radiation during the hadron era. Our matter, that actually exists right now, will therefore be composed half and half of matter and antimatter and will owe its existence to pure thermal radiation. This is "Let there be Light" in all its simplicity. OMNÈS estimates that above a critical temperature $T_c \simeq 3 \cdot 10^{12}$ K there is a transition phase: regions very rich in nucleons N and other regions very rich in antinucleons $\bar{N}$ form, this phenomenon arising essentially because of the appearance of a statistical repulsion force N-$\bar{N}$ much stronger than the dynamic repulsion N-N or $\bar{N}$-$\bar{N}$. A superficial pressure will also appear, tending to increase the contrast between these regions, which form a suspension of matter within antimatter, or vice versa. At the contact of two regions made of opposite types of matter annihilation produces light particles which, because of their vigorous motion, create a pressure that tends to separate the two regions. This is the annihilation pressure of ALFVEN and KLEIN, creating an effect analogous to calefaction. It is thanks to this effect that all the matter-antimatter did not entirely disappear from the universe.

Towards the end of the lepton era we find a plasma of electrons and photons in which there is, at places, a sparse scattering of some nucleons or antinucleons. At the periphery of such regions annihilation continues, and its pressure causes these regions to coalesce into larger and larger zones which will eventually give birth to galaxies or antigalaxies.

The discovery of the microwave 3 K radiation has given rise, as we saw, to compelling research of a new type in which particle physics and statistical mechanics are united in the exploration of an absolutely exceptional domain: cosmology in the first ten-thousandth of a second.

The ultimate theory has been proposed by TRYON: the universe could be a fluctuation of the vacuum, in the sense of quantum mechanics. This idea is based on the fact that the gravitational energy of a mass m interacting with a volume of radius c/H and of density ρ_{EdS} (10.53) is $-mc^2$, practically opposed to its mass energy. To within a factor close to unity the total energy of the universe may therefore be zero. If, additionally, we suppose that the universe contains as much matter as antimatter, it may effectively be a fluctuation of the vacuum. In the same way that an electron-positron pair can seem to materialise from the vacuum with energy ΔE during a mean time Δt, such that $\Delta E \cdot \Delta t = h$, the universe also has a non-zero probability of existing for 10^{10} years with its observed energy. And as we are here to verify these things, this tiny probability that the universe, by chance, came out of the vacuum, without violating the laws of physics, *has* happened at least once...

Numerical Constants

$\pi = 3.14 \quad e = 2.72 \quad \log_e 2 = 0.693$
Velocity of light, $c = 3.00 \cdot 10^{10}$ cm s^{-1}
Gravitational constant, $G = 6.67 \cdot 10^{-8}$ dyne cm^2 g^{-2}
Planck constant, $h = 6.63 \cdot 10^{-27}$ erg s
Boltzmann constant, $k = 1.38 \cdot 10^{-16}$ erg deg^{-1}
Stefan's constant, $a_s = 7.56 \cdot 10^{-15}$ erg cm^{-3} deg^{-4}
Electron charge $= 4.8 \cdot 10^{-10}$ cgs units
Electron mass $= 9.11 \cdot 10^{-28}$ g
Proton mass $= 1.67 \cdot 10^{-24}$ g
Mass of the Sun $= 1.99 \cdot 10^{33}$ g
1 parsec (pc) $= 3.08 \cdot 10^{18}$ cm $= 3.26$ light years
1 year $= 3.16 \cdot 10^7$ s
1 electron volt (eV) $= 1.6 \cdot 10^{-12}$ erg
Intrinsic bolometric luminosity of the Sun $= 3.90 \cdot 10^{33}$ erg s^{-1}
Magnitudes: Expression of luminosity on a logarithmic scale in which
the absolute magnitude of an object of intrinsic luminosity L is given
by $M = \mathcal{M}_\odot - 2.5 \log_{10}(L/L_\odot)$ where $L_\odot$ is the intrinsic luminosity of
the Sun and $\mathcal{M}_\odot$ its absolute magnitude. $\mathcal{M}_\odot$ (bolometric) = 4.72;
$\mathcal{M}_\odot$ (photographic) = 5.37; $\mathcal{M}_\odot$(B) = 5.41. The apparent magnitude of
an object with absolute magnitude M having a distance modulus μ is,
in the absence of absorption, $m = M + \mu$.

Bibliography

PART I:

Abd-al-rahman al-Sûfi: *Description des étoiles fixes*, traducted by H.C.F.C. Schjellerup (964)
Sandage, A. & M., Kristian, J. (eds.): *Galaxies and the Universe* (Univ. of Chicago Press, 1975)
Balkowski, C., Westerlund, B.E. (eds.): *Decalages vers le rouge et expansion de l'univers. L'évolution des galaxies et ses implications cosmologiques*; (CNRS, 1977)
De Vaucouleurs, G. & A., Corwin, H.G., Jr.: *Second Reference Catalogue of Bright Galaxies* (Univ. of Texas Press, 1977)
De Vaucouleurs, G.: The extragalactic distance scale. Astrophys. J., 223 (1978)
Sandage, A., Tammann, G.A.: Steps towards the Hubble constant. Astrophys. J., 190 (1974)
Longair, M.S., Einasto, J.E. (eds.): *The large Scale Structure of the Universe* (D. Reidel, 1978)
White, R.A.: Morphological characteristics of clusters of galaxies. Astrophys. J., 226 (1978)
Rood, H.J.: Nearby groups of galaxy clusters. Astrophys. J., 207 (1976)
Gunn, J.E., Longair, M.S., Rees, M.J.: *Observational Cosmology* (Saas-Fee Course, Geneva Observatory, 1978)
Rubin, V.C., Thonnard, N., Ford, W.K., Jr.: Motion of the Galaxy and the Local Group. Astron. J., 81 (1976)
Tinsley, B.M.: Evolutionary synthesis of the stellar population in elliptical galaxies. Astrophys. J., 222 (1978)
Visvanathan, N.: Distance to the Virgo cluster. Astron. Astrophys., 67 (1978)
Haynes, M.P., Brown, R.L., Roberts, M.S.: A search for atomic hydrogen in clusters of galaxies. Astrophys. J., 221 (1978)
De Young, D.S., Roberts, M.S.: The stability of galaxy clusters. Astrophys. J., 189 (1974)
Forman, W. et al.: The detection of large X-ray halos in clusters. Astrophys. J., 225 (1978)
Lake, G., Partridge, R.B.: Detection of intergalactic gas in clusters. Nature, 270 (1977)
Berkhuijsen, E.M., Wielebinski, R. (eds.): *Structure and Properties of Nearby Galaxies* (D. Reidel, 1978)
Jauncey, D.L. (ed.): *Radio Astronomy and Cosmology* (D. Reidel, 1977)
Longair, M.S. (ed.): *Confrontation of Cosmological Theories with Observational Data* (D. Reidel, 1974)
Spinrad, H. et al.: Halos of spiral galaxies. Astrophys. J., 225 (1978)

PART II:

Lichnerowicz, A.: *Eléments de calcul tensoriel* (Armand Colin, 1960)
Cartan, E.: *Leçons sur la géométrie des espaces de Riemann* (Gauthier-Villars 1946)
Peschl, E.: *Über die Gestalten des Raumes* (Münster Universität, 1935)
Rinow, W.: *Die Innere Geometrie der Metrischen Räume* (Springer, 1961)
Lelong-Ferrand, J.: *Géométrie différentielle* (Masson, 1963)

PART III:

Robertson, H.P., Noonan, T.W.: *Relativity and Cosmology* (W.B. Saunders Co., 1968)
McVittie, G.C.: *General Relativity and Cosmology* (Chapman and Hall Ltd., 1965)
Sandage, A.: The ability of the 200-inch telescope to discriminate between world models. Astrophys. J., 133 (1961)
Sandage, A.: The light travel time of distant galaxies. Astrophys.J., 134 (1961)
Sandage, A.: The change of redshift and apparent luminosity of galaxies. Astrophys. J., 136 (1962)
Refsdal, S., Stabell R., de Lange, F.G.: Numerical calculations on relativistic cosmological models. Mem. Roy. Astron. Soc., 71 (1967)
Rindler, W.: Visual horizons in world-models. Mon. Not. Roy. Astron. Soc., 116 (1956)
Campusano, L., Heidmann, J., Nieto, J.L.: The age of the universe in Friedmann models. Astron. Astrophys., 41 (1975)

PART IV:

Selected works cited in Part I plus:

Kristian, J., Sandage, A., Westphal, J.A.: The extension of the Hubble diagram. Astrophys. J., 221 (1978)
Coleman, G. et al.: The spectrum of the quasar PHL 957. Astrophys. J., 207 (1976)
Giacconi, R., Gursky, H. (eds.): *X-Ray Astronomy* (D. Reidel, 1974)
Fichtel, C.E., Simpson, G.A., Thompson, D.J.: Diffuse gamma radiation. Astrophys. J., 222 (1978)
Swanenburg, B.N.: Observation of high energy gamma-radiation from 3C 273. Nature, 275 (1978)
Stockton, A.: The nature of quasar redshifts. Astrophys. J., 223 (1978)
Smoot, G., Gorenstein, M.V., Muller, R.A.: Detection of anisotropy in the cosmic blackbody radiation. Phys. Rev. Lett., 39 (1977)
Kaufmann, M., Thuan, T.X.: Young massive galactic halos at large redshifts. Astrophys. J., 215 (1977)
Sunyaev, R.A., Tinsley, B.M., Meier, D.L.: Observable properties of primeval giant elliptical galaxies. Comments Astrophys., 7 (1978)
Casini, C., Heidmann, J., Tarenghi, M.: Investigation of the clumpy irregular galaxy Markarian 296. Astron. Astrophys., 73 (1979)
Steigman, G.: Observational tests of antimatter cosmologies. Annu. Rev. Astron. Astrophys., 14 (1976)
Brout, R., Englert, F., Gunzig, E.: The creation of the universe as a quantum phenomenon. Ann. Phys., 115 (1978)

Subject Index

Absolute 73
  bolometric magnitude 21,114,161
  differential 58
  horizon 131
  magnitude 3,114,161
Age of universe 40,110,125
Antenna gain 28
Antimatter 159
Antipole 78,82,133
Apparent bolometric magnitude 21,113, 161
Axiom of the plane 62

Beam efficiency 29
  of antenna 26
Black body 152
  holes 37
Bolometric brightness 115
Bremsstrahlung 31
Brightness temperature 26

Calefaction 159
Cayley distance 73
Cellular structure 18
Central representation 76
Cepheid 3
Christoffel symbols 44
Clumpy irregular galaxy 156
Cluster counts 15
  of clusters 15
  of galaxies 11
Coconcentration 116

Codensity 100
Comoving coordinates 98
Compton wavelength 157
Conformal representation 78
Continuum 26
Contracting a tensor 68
Contravariant 43
Coordinate line 43
Cosmic repulsion 123
  time 99
Cosmological constant 93,119
  horizon 127
Covariant 44
  derivative 58
Covolume 100
Curvature tensor 67

Deceleration parameter 102
Density of the universe 37
  parameter 103,120
De Sitter model 123
Development 46,48,59
  angle 101
Diameter 110, 147
Dingle equations 94
Distance indicator 3
  modulus 3,114
Distribution of radio luminosity 145
Divergence 59

Eddington-Lemaitre model 122

Einstein-de Sitter model 101, 122
  -Eddington model 122
  equations 93
  model 122
Elliptic space 69
Emden equation 34
Energy-momentum tensor 90
Equation of state 99
Equator 78,130,133
Equatorial surface 78,82
Euclidean contact space 59
Event 129
  horizon 130
Evolution of quasars 147
  of radiogalaxies 147
  of the universe 154
Expansion 19,98,107

First order representation 57
Fluctuation of the vacuum 160
Flux 112
  densitiy 114,143
Formation of galaxies 156
Four-velocity 90
Free mobility 68
Friedmann model 100
Fundamental polyhedron 50

Galactic rotation 20
Galaxy counts 12,17
Gauss' theorem 67
Generating operation 49
Geodesic curve 59
  representation 64,73
  surface 62
Gigaparsec 2
Gradient 17
Gravitational potentials 93
Group of galaxies 7,10

HII regions 4

Hadron era 156,159
Helium abundance 40,155
Hidden mass 36,38
Hierarchical universe 37
Holonomy group 49
Homogeneous model universe 97
Homologous point 49
Hubble constant 21,24,102
  diagram 21,125,136
  law 21,107
  time 24
Hyperbolic space 73,76

Intergalactic absorption 25
  dust 25
  gas 31
Inverse Compton effect 142
Ionized Hydrogen 140
Isotropic space 67,94
Isotropy 94,97,141,148,152

K-correction 21
Klein bottle 55

Lemaitre model 122
Lepton era 159
Light cone 91,92
Light year 161
Limiting cone 130
Local Group 9
  reference frame 20
  Supercluster 16
Locally Eudlidean space 46
  hyperbolic space 65,71
  non-Euclidean space 65
  spherical space 65,69
Lowering an index 45
Luminosity class 5
  function 9,12,13,15,35
Lyman α line 139

Mattig's formula 111
Metagalaxy 2
Metric element 43
   tensor 43,57
Milne model 123
Minkowski spacetime 89
Moebius space 54
Molecular hydrogen 140
Monochromatic bolometric brightness 115

Natural frame 44
Neutral hydrogen 26
   hydrogen mass 27
Neutrino 36
Non-Euclidean space 72
Normal representation 80
Number counts 115

Optical depth 28,139,140,141
Osculating Euclidean space 58

Palomar Sky Survey 5
Parallelism 59
Parallel transport 60
Parsec 161
Particle horizon 127
Pavement of space 50,69
Perfect fluid 90,95
Planck formula 152
   length 157
Pole 78
Primordial galaxies 156
Principle of geodesics 93
Proper density 90
   time 98

Quantum era 157,159
Quarks 157
Quasar 143
   counts 147

Radiant power 113
Radiogalaxy 143
Radio luminosity 145
   luminosity function 145
   magnitude 146
   sources counts 144
Radius of curvature 77
Raising an index 85
Random motion of galaxies 21
Rayleigh formula 26
Reduced cosmological constant 120
   sinusoidal representation 80
Regular Riemannian space 57
Ricci tensor 68
Rich cluster 14
Riemann-Christoffel tensor 46,65
Riemannian curvature 69
Robertson-Walker metric 97
Rotation velocity 7

Scalar curvature 68
Scaling parameter 101
Schur's theorems 62,64
Second order representation 58
Seyfert galaxies 141
Signature 89
Sinusoidal representation 80
Spacelike line 91
Spacetime 89
Spectral index 115
   ratio 107
   shift 20,107
Spectrum 114
Spherical space 73,77
Spin temperature 28,29
Steady state theory 148
Stefan's law 104
Summation of indices 43
Supercluster 15
Supergalactic plane 16
Supergalaxy 16

Superspace   158

Tangent Euclidean space   58
Thompson scattering   31,141
Timelike line   91
Totally geodesic surface   62
Travel time   108
21-cm flux   27
    line   26

Uniform model universe   98
Unitary velocity   90

Virial theorem   36

Worldline   90

X- and γ-radiation   31,141
X-ray clusters   31
    halos   36

W. Högner, N. Richter

Isophotometric Atlas of Comets, Part 1

1980. 90 plates, comments and tables.
ISBN 3-540-09171-8

Isophotometric Atlas of Comets, Part 2

1980. 55 plates, comments and tables.
ISBN 3-540-09172-6

Springer-Verlag
Berlin
Heidelberg
New York

This beautiful atlas contains a carefully selected collection of material needed for the study of the physics of comets. The authors scrutinized more than 300 photographs taken in the years 1902–1967. They applied photographic equidensities as quasiisophotes according to the method of Lau and Krug by using the Sabattier effect. Reproductions of the original photographs, their isophote diagrams as well as enlarged isophote diagrams of the cometary heads and nuclei are presented. The IAU considers such an atlas to be of "extreme value" to the astronomical community.

K. R. Lang

Astrophysical Formulae

A Compendium for the Physicist and Astrophysicist

2nd corr. and enlarged ed. 1980. 46 figures, 69 tables. XXIX, 783 pages
ISBN 3-540-09933-6

Astronomy and Astrophysics Abstracts

A Publication of the Astronomisches Rechen-Institut Heidelberg Member of the Abstracting Board of the International Council of Scientific Unions

Volume 25
Literature 1979, Part 1
Editors: S. Böhme, U. Esser, W. Fricke, I. Heinrich, W. Hofmann, D. Krahn, D. Rosa, L. D. Schmadel, G. Zech
1979. X, 871 pages
ISBN 3-540-09831-3

Volume 26
Literature 1979, Part 2
1980. X, 794 pages
ISBN 3-540-10134-9

Springer-Verlag
Berlin
Heidelberg
New York